APPLIED MICROECONOMICS

Problems in Estimation, Forecasting, and Decision-Making; Student's Manual

MAVEN BOOKS

APPLIED MICROECONOMICS

Problems in Estimation, Forecasting, and Decision-Making; Student's Manual

RICHARD L. SCHMALENSEE

MAVEN BOOKS

Chennai Trichy Tirunelveli New Delhi

MAVEN BOOKS

An Imprint of **MJP Publishers**

ISBN 978-93-88191-16-6 **MAVEN Books**

No. 44, Nallathambi Street, Triplicane, Chennai 600 005

Printed and bound in India

MJP 590

Publisher : **C. Janarthanan**

PUBLISHER'S NOTE

The legacy of a country is in its varied cultural heritage, historical literature, developments in the field of economy and science. The top nations in the world are competing in the field of science, economy and literature. This vast legacy has to be conserved and documented so that it can be bestowed to the future generation. The knowledge of this legacy is slowly getting perished in the present generation due to lack of documentation.

Keeping this in mind, the concern with retrospective acquiring of rare books has been accented recently by the burgeoning reprint industry. MAVEN Books is gratified to retrieve the rare collections with a view to bring back those books that were landmarks in their time.

In this effort, a series of rare books would be republished under the banner, "MAVEN Books". The books in the reprint series have been carefully selected for their contemporary usefulness as well as their historical importance within the intellectual. We reconstruct the book with slight enhancements made for better presentation, without affecting the contents of the original edition.

Most of the works selected for republishing covers a huge range of subjects, from history to anthropology. We believe this reprint edition will be a service to the numerous researchers and practitioners active in this fascinating field. We allow readers to experience the wonder of peering into a scholarly work of the highest order and seminal significance.

MAVEN Books

CONTENTS

CHAPTER I

Introduction

This manual describes a sequence of exercises in applied microeconomics designed to supplement intermediate level courses in microeconomic theory or managerial economics. A working knowledge of elementary calculus and the availability of a standard computer program for multiple regression are assumed. The tools and concepts needed for these exercises are generally discussed herein, though the treatment of most topics is not intended to be complete. Throughout, the focus is on understanding rather than rigor.

The problems presented here are, of course, simpler than most encountered in the real world. While they should present a fair challenge, they are not intended to convey all of the difficulties encountered in actual applied work. Rather, the focus is on learning to use powerful tools and concepts in simple, but non-trivial situations. These same approaches can be applied to more complex problems, though not as easily as they can be used here. The hope is that working on fairly simple problems will make it easier for the student when he is faced with complicated situations of the same basic type.

The first three exercises involve analyzing historical data for a particular industry. In the industry to be analyzed, a large number of firms operated in the period for which data will be provided. There was no technical change in this period, and no major shifts in consumers' tastes occurred. The output of each firm in the industry was either determined directly by the government or fixed by arbitrary non-price rationing of an essential input factor. As we will see in the next chapter, this means that simple yet powerful statistical methods can be used to analyze the data. Chapter II will describe and discuss these methods.

Chapter III outlines the first two exercises and describes the history of the industry in more detail. From data provided in class, you will be required to estimate the demand and cost relationships that characterize the market to which you have been assigned. The instructor will provide directions on the use of the particular regression program to be employed. Chapter III concludes with a description of the two short reports you should write on these exercises.

Chapter IV outlines the theory of price determination in competitive and monopolistic markets and discusses the simple mathematics it is based on. You are to apply this theory to your industry. For given values of certain variables, you will forecast the price and output of the industry you are dealing with under conditions of pure competition and of perfect monopoly.

The next chapter examines price determination in oligopolistic markets, markets with few firms. Under these circumstances, firms' expectations about their competitors' behavior become an important determinant of industry performance. This complicates the firms' decisions to the point where no truly general theories of oligopolistic pricing have been developed.

To give students a deeper understanding of the problem of pricing in concentrated markets than can be provided by passively reading about it, the last exercise involves making pricing decisions under conditions of oligopoly. Chapter VII describes a simulation exercise in which each student will be on the board of directors of a firm with few competitors. Profits of each firm will depend on the prices of all firms in the same market, and individual's grades will depend to some extent on the profits of their firms. At the end of the exercise, the behavior and performance of the firms and industries simulated are to be analyzed in terms of their structure and the theoretical notions outlined in Chapter VI.

All four of these exercises are related. The statistical estimations should provide you with some idea of the cost function of a typical one-plant firm and of the demand curve faced by that firm when all producers charge the same price. In the forecasting problem, you will be given the number of firms operating in the industry during the period for which data were provided. This will enable you to construct an estimate of the industry's demand function and to forecast short-run and long-run competitive and monopoly price. A good estimate of the short-run monpoly price will be extremely useful in the decision-making exercise, since if all firms in an industry charge this price they will maximize total industry profits. In this exercise the demand function you originally estimated will be the curve facing all firms if they all charge the same price. When prices differ within an industry, however, you will have to learn by experience the impact of price differentials on firms' demands.

Finally, none of the four exercises will necessarily be the same from class to class. The instructor can provide data corresponding to a large number of different cost and demand functions, and the market structures encountered in the decision-making exercise can vary considerably from industry to industry.

CHAPTER II

Econometric Methods

What is Econometrics?

Economics attempts to be a science, and therefore economists are interested in explaining and predicting certain phenomena. Economic theory enables us to make some predictions, but these normally relate only to directions of change. We expect, for instance, that if a monpolist lowers his price his sales will increase. Sometimes even this vague sort of prediction is denied us. No pure theory exists which will tell us whether GNP will rise or fall next year.

Economic theory does point to the existence of certain fairly stable relationships involving quantities of interest. These are often capable of being expressed in equation form. To make predictions and decisions, we need to know what those equations look like; we need numbers. This need gave rise to econometrics, which is often defined as the measurement of relationships suggested by economic theory. Econometrics also contains a set of techniques that enable us to test propositions suggested by economic theory, as we will see below.

Economic theory emphasizes the interdependence of economic systems. Any one quantity, such as the price of ice cream, is influenced by a virtually infinite number of other quantities. If we wish to construct a demand curve for ice cream that is actually useable, though, we can't consider all possible variables. We must make a judgement about which are important. We might, for instance, decide that the main variables influencing the demand for ice cream are the weather, the price of ice cream, the general level of other prices, consumers' incomes, the age distribution of the population, and advertising of ice cream. We then group all other influences into an error or disturbance term. Since there are so many forces that have been left out, and since each one is so small, we treat the error term as a random variable, a chance-determined quantity whose values cannot be predicted. From the way we have defined it, the error term cannot be observed.

Formally, then, we begain with equations of the following form:

$$Y = f(X_1,\ldots,X_n,e) \tag{2.1}$$

The dependent variable is Y, and the quantities $X_1,\ldots,X_n$ are the principal independent variables influencing Y. Other forces are summarized in the error term, e. This equation is assumed to hold for all observations on Y and the X's. (An observation may be the values taken on by the variables in a particular year or other time-period, or the data may be composed of observations on individual states or households or other units in a single time-period. These two types of observations are termed time-series and cross-section, respectively). Given observations on the X's and Y, the task of econometrics is to describe the function $f(\cdot)$ as well as possible.

Economic theory may tell us something about the function f(•). Often the signs of the partial derivatives are predicted by the theory; we expect the quantity of icecream demanded to decline as the price of ice cream rises. Sometimes other information is available as well. Consumer demand theory, for instance, says that if all prices and income double, a consumer's purchases should remain unchanged. All information the theory provides should be incorporated into econometric analysis. Rarely does this information completely determine the form of the function f(•).

For reasons of simplicity, whenever it is sensible, we work with functions that are linear in a finite set of unknown parameters. That is we deal with special cases of (2.1) of the form

$$g(Y) = b_1 h_1(X_1) + \dots + b_n h_n(X_n) + e. \tag{2.2}$$

The b's are the unknown parameters that it is desired to estimate. The functions g(•) and $h_1(\cdot) - h_n(\cdot)$ are specified by the econometrician. Special cases of (2.2) are, for instance, the following:

$$Y = b_1 X_1 + b_2 X_2 + e$$

$$\log (Y) = b \log (X_1) + b_2 \log (X_2) + e$$

$$Y = b_1 + b_2 X_1 + b_3 (X_1)^2 + e$$

In the third of these examples, b_1 is multiplied by a variable which is always equal to one. Thus the model has an intercept or constant term. Only rarely does economic theory greatly restrict the choice of functions in (2.2). When some information can be distilled it should be used. In general, though, g(•) and the h(•)'s should be chosen so as to be moderately simple and easy to work with and so as to explain the data as well as possible. The information in the data can be used to select functional forms.

The most common technique employed to estimate the b's in an equation of the form of (2.2) is called ordinary least squares regression or multiple linear regression. The main purpose of this chapter is to provide the reader with an elementary, intuitive understanding of the use of this tool. First, however, we must consider a situation in which the use of least squares is not appropriate. This is when our model of an economic phenomenon involves variables which are mutually determined, though a system of simultaneous equations. This situation is considered in the next section. We then turn to some essential statistical notions. The last two sections of this chapter discuss linear regression, first in statistical terms and then from the point of view of a user wishing to interpret computer output from a regression program. Our treatment throughout is sketchy and non-rigorous; the references listed at the end of the section are recommended for filling the gaps we leave.

Simultaneous Equations

Consider a supply - demand system where P is price, Q is quantity, and the true equations are the following:

$$P = b_1 + b_2Q + e \qquad \text{supply curve}$$

(2.3)

$$Q = b_3 - b_4P + u \qquad \text{demand curve}$$

Here e and u are error terms, and the b_i are constants. Suppose you were given data on this market for a long period of time. Could you estimate the supply and demand curve? The answer is clearly no, since all the observations on P and Q would be clustered around the equilibrium point. If you knew that e was generally much larger than u in absolute value, you would know that the supply curve moved much more than the demand curve. Most of the data would lie near the demand curve then, and you could get so idea of its shape. But there would be no possible way to use the data to estimate the parameters b_1 and b_2 of the supply curve.

In a classic early study of the demand for agricultural commodities, this notion was employed. The argument was that supply varied a great deal depending on the weather, but that little random movement in the demand curve was to be expected. In this situation, we say that the demand curve is <u>identified</u>, that is, it is potentially capable of being estimated from the data, while the supply curve is not identified (or under-identified).

Suppose now that u and e in (2.3) have about the same average (absolute) size, bu t that demand depends on the level of income as well as price. Then changes in income will move the demand curve and trace out the supply curve. The supply curve is identified because o f the variable exogenous to (determined outside of) the system of equations we are dealing with. If supply also depend on rainfall, and if rainfall can be measured, it can be shown that both curves are identified. The references at the end of the Chapter discuss the identifications problem in more detail. The main idea is that exogenous variables must be present in some equations and not in others, or identification is impossible.

Suppose that (2.3) is expanded to the following system:

$$P = b_1 + b_2Q + b_5R + e \qquad \text{supply curve}$$

(2.4)

$$Q = b_3 - b_4P + b_6Y + u \qquad \text{demand curve}$$

Here b_5 and b_6 are constants, R is rainfall and Y is consumer income. As we mentioned just above, these equations are identified. Can we use simple linear regression to estimate the unknown constants? The answer is generally no.

As written above, (2.4) is composed of structural equations, equations suggested by the theory. If you solve (2.4) for P and Q, you obtain two reduced form equations in which the endogenous variables P and Q are functions of the exogenous variables R and Y and the disturbance terms e and u. Since both P and Q depend on both e and u, it is clear that e and Q will move together in the first equation. If e and Q are correlated in this fashion, least squares will give results that are incorrect (inconsistent in the statistician's terminology) and generally misleading.

The intuitive nature of the problem is moderately simple. Suppose e is large whenever Q is. We cannot observe e, though. Thus the effects of both e and Q on P will be attributed to Q, and any method of estimating b_2 that does not take this into account is likely to come up with an estimate well above the true value.

Figure II.1 illustrates this discussion. We assume that data are generated according to the equation

$$Y = b_0 + b_1 X + e,$$

where e is, as usual, an error term. This equation, with e = 0 is shown as the True Line in the Figure. Assume that e is positively related to X: whenever e is large, X will tend to be large also. In this case, the data will be as shown in the Figure, where each "o" represents an observation of X and Y. Looking only at the data, one would take the Estimated Line as a good estimate of the (unknown) True Line. But such as estimate would clearly be misleading. Formally, the estimated parameters are biased and inconsistent; no matter how many observations we obtained, our estimates would still differ systematically from the true values of b_0 and b_1

Figure II.1

Illustration of Biased Estimates

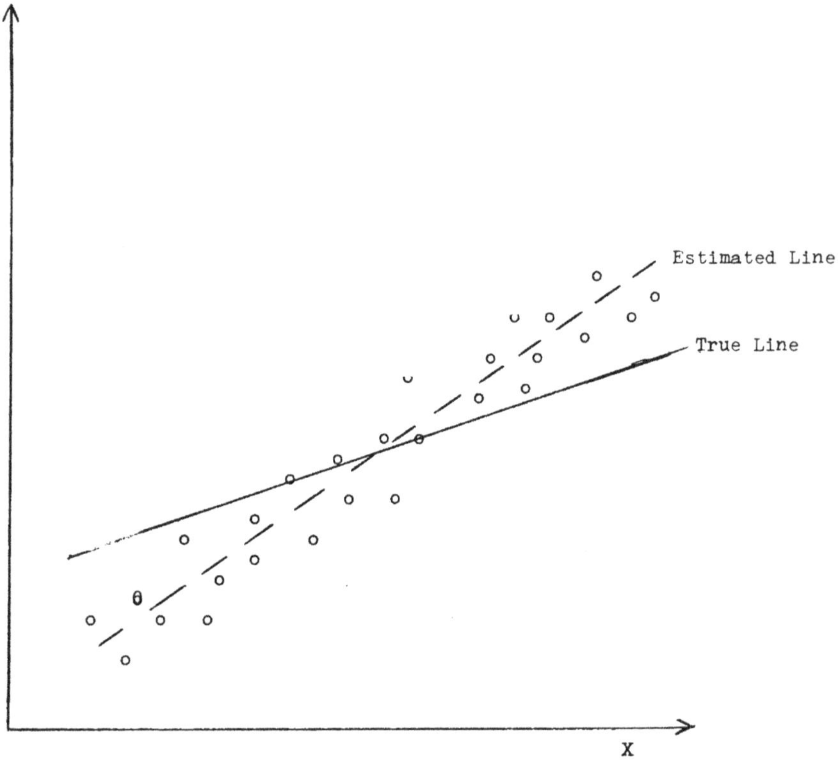

Simultaneous equation models thus have two sorts of problems. The first is identification: is it theoretically possible to "sort out" the coefficients of the structural equations from observable data generated by the true equation system? The second problem is estimation: if least squares gives incorrect estimates, what can be used? These problems are discussed in the references.

Basic Statistical Concepts

Before we can discuss least squares regression, we must venture briefly into the realm of statistical analysis. We have said that we will treat the disturbance or error term in econometric equations as a random variable. As such, its particular values cannot be predicted. But some useful things can be said about random variables, and we shall proceeed directly to say a few of them. The results of this section are quite essential for the rest of the chapter; the sensible use of regression depends on an understanding of the properties of random variables that we shall develop here.

Suppose we have a black box with a crank. Every time the crank is turned, a poker chip is ejected. On each poker chip is a number. The numbers cannot be predicted. Each number is an observation of a random variable we will call x. Suppose we take the box apart. By definition, this will not enable us to predict individual values of x. But we will be able to summarize the mechanism generating these values of a density function, $f_x(\cdot)$, which is defined by the following equation:

$$(2.5) \qquad \text{prob}\ (a < x \leq b) = \int_a^b f_x(x)dx$$

In words, the probability that a particular value of x will lie between a and b is the area under the density function between a and b. The density function must always be non-negative to make any sense, since probabilities cannot be less than zero. Also, the integral of $f_x(x)$ from $-\infty$ to $+\infty$ must be equal to one, since some number is on each poker chip.

Remember that we know the density function only because we have dismantled the black box. In the real world, we will never know for sure the exact form of any density function we are interested in.

The next main concept is that of expected value. The expected value of some function g(x) of the random variable x is written as E[g(x)] and is defined by

$$(2.6) \qquad E[g(x)] = \int_{-\infty}^{+\infty} g(x)\ f_x(x)dx.$$

The expected value is sort of weighted average, where the weights are probabilities of occurance. Two common and useful expected values are the following:

$$E[x] = \mu_x = \text{the mean of } x$$

$$E[(x - \mu_x)^2] = \sigma_x^2 = \text{the variance of } x$$

The mean of x is a measure of the central tendancy of the distribution, an indication of the average value of x. The variance, on the other hand, measuresthe spread or dispersion of x around its mean. The square root of the variance, written σ_x, is called the standard deviation of x. Expected values of the form $E[x^k]$ or $E[(x - \mu_x)^k]$, where k is a positive integer, are often called moments of the probability distribution of x. The mean at x is thus the first moment, while the variance is the second moment about the mean.

We need two further definitions. Imagine another black box that produces poker chips with two numbers on them, one labeled x and the other labeled y. Dismatleing the box would give use enough information to write down the joint density function of the two random variables. This function is defined by

$$(2.7) \qquad \text{prob } (a < x \leq b \ \underline{\text{and}} \ c < y \leq d) = \int_a^b \int_c^d f_{xy}(x,y)dy\,dx.$$

The natural generalization of (2.6) enables us to talk of the expected value of functions of x and y.

Two of these expected values are of great interest to us. The definitions are

$$E[(x - \mu_x)(y - \mu_y)] = \text{cov}\,(x,y) = \text{the covariance of } x \text{ and } y$$

$$\frac{\text{cov}(x,y)}{\sigma_x \sigma_y} = \rho(x,y) = \text{the correlation coefficient of } x \text{ and } y.$$

The covariance is a measure of association. If both variables are always above and below their means at the same time, the covariance will be positive. If x is above its mean when y is below its mean and vice versa, the covariance will be negative. The covariance may have any value, depending on the size of x and y. If x were measured in inches instead of feet, for instance, the covariance of x with any other random variable would be multiplied by twelve. The correlation coefficient is a measure of association that does not suffer from this defect. It can be shown that the correlation coefficient will always be between -1 and +1.

As we mentioned, nature very rarely delivers the exact form of density functions. Suppose we have T observations on x, call them $x_1, \ldots, x_T$. What can these twli us about the density function that generated them? One approach might be to estimate the moments of the density function. The true moments involved averaging over the density function, weighting by the true or population probabilities. We can estimate moments by averaging over the observed or sample probabilities. We could estimate the true mean of x, for instance, by the sample mean:

$$\hat{\mu}_x = \frac{1}{T} \sum_{i=1}^{T} x_i .$$

Throughout this chapter, a hat (^) denotes an estimated value. The population variance of x could be similarly estimated from our data by the sample variance:

$$\hat{\sigma}_x^2 = \frac{1}{T} \sum_{i=1}^{T} (x_i - \mu_x)^2$$

Estimates like these will differ from sample to sample: the value taken on in any one sample is a random variable, since it is a function of random variables. This notion should not be too difficult. Throw a die ten times. Count the number of times it comes up six, and divide this number by ten. This is an estimate of a probability. It is a random variable, though, since you cannot predict the estimate you will obtain if you throw the die another ten times.

This may begin to sound relevant to econometrics. In working with a model of the form of (2.2), we assume that there are true, underlying, population values of the b's that lead to the data we have collected. We want to obtain a set of estimates of them for use in decision-making. These estimates will depend on the exact values taken on by the (unobservable) disturbance term in the sample we have. Since our estimates of the b's depend on the random variable e, they are also random variables. In order to say anything about these estimates, for instance to decide how precise they are, we need to assume something about the density function of the error term.

The most common assumption and the most useful one is that the error term is normally distributed. The density function of a normally distributed random variable x is given by

$$f_x(x) = \frac{1}{\sigma_x \sqrt{2\pi}} \quad e^{-1/2(x - \mu_x)^2 / \sigma_x^2}$$

Notice that this function depend only on two parameters: the mean and variance (or standard deviation). If a random variable is thought to be normally distributed, an estimate of its density function can be written down as soon as estimates of the mean and variance are calculated.

Figure II.2 shows graphs of the normal density function for various values of the parameters. (None of the functions are actually touching the horizontal axis, in spite of my artwork.) It should be clear that the mean indicates cnetral tendancy, while the variance measure dispersion.

The rationale for using the normal distribution in econometrics lies in the so-called Central Limit Theorem. This theorem says, roughly, that the sum of a very large number of independently distributed, very small random variables is normally distributed. This is true, as the name of the theorem implies, only in the limit as the number of small variables approaches infinity. But the approximation is quite good for "moderate" numbers of variables. This relates to our earlier discussion of the error or disturbance term as the net result of a large number of small forces, so the assumption of a normal distribution

Figure II.2

Comparison of Different Normal Density Functions

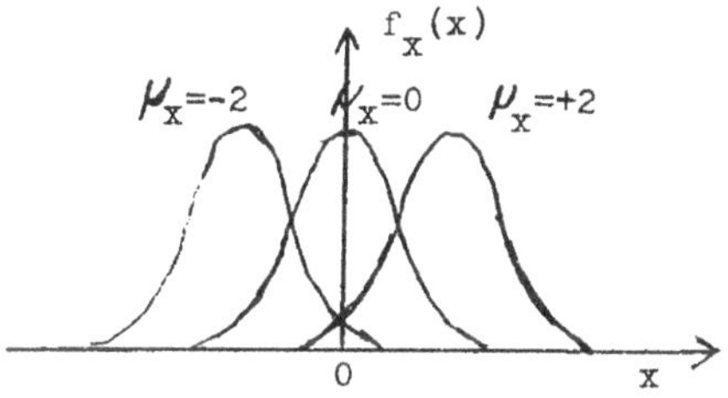

(a) different means, same variance

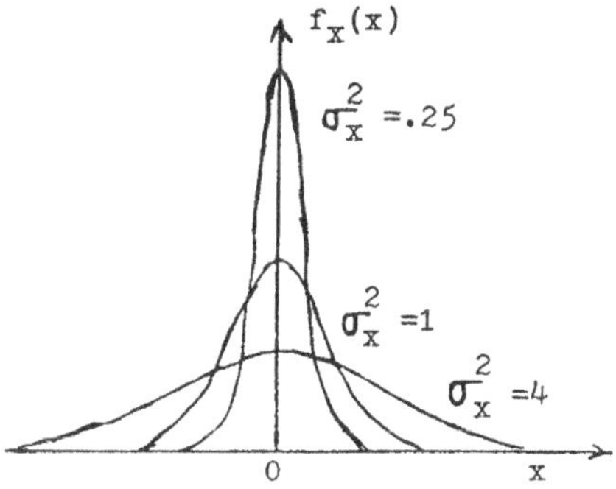

(b) different variances, same mean

is quite natural. Least squares is a good estimator even if this assumption is wrong, and the testing procedures we will discuss are fairly reliable if the assumption is even approximately correct.

Two useful properties of normally distributed variables are the basis for the tests we will be concerned with in our discussion of least squares:

$$\text{prob}[|x - \mu_x| > 2.0\ \sigma_x] \simeq .05$$

$$\text{prob}[|x - \mu_x| > 2.6\ \sigma_x] \simeq .01$$

In words, the probability that a single observation of a normally distributed random variable x will be two standard deviations away from the mean is about .05. The probability that it will be 2.6 standard deviations away is about. .01.

Suppose we know the variance of a normally distributed random variable, but we do not know its mean. We have T observations on the variable, and we compute the sample mean, $\hat{\mu}_x$, defined above. What does this estimate tell us about the true mean? It is easy to show that if a random variable x, not necessarily normally distributed, has a standard deviation σ_x, the quantity $\hat{\mu}_x$, which is also a random variable, will have standard deviation σ_x / T

If x is normally distributed with mean μ_x, so is $\hat{\mu}_x$.

Suppose that we have T=25 observations on a particular variable x with known standard deviation equal to five. Thus the quantity μ_x is normally distributed with standard deviation one. Suppose the sample mean is equal to five. If the population mean were 1.9, the probability that we would observe a sample mean of five is less than .01, since five is more than 2.6 standard deviations away from 1.9. Now the population mean is a fixed number, and we can't really say that it is unlikely to be equal to 1.9. It either is 1.9 or it isn't. What we can say is that it is very unlikely that we would have obtained the observations we actually got if the population mean is outside the interval 2.4 - 7.6. This is a key notion, and it underlies the theory of hypothesis testing and confidence intervals.

We have just seen an example of the latter. We expressed confidence in the above example that the true population mean in the interval 2.4 - 7.6. In fact, we associate a confidence coefficient of .99 with this interval. With .95 confidence, we can assert that it is unlikely that a population mean outside the range 3.0 - 7.0 would give rise to our data. In general, an approximate 95% confidence interval on the mean of a normal distribution with known variance is given by

$$\hat{\mu}_x - 2\sigma_x/\sqrt{T} < \mu_x < \hat{\mu}_x + 2\sigma_x/\sqrt{T}.$$

Procedures for constructing confidence intervals on unknown parameters are closely related to the methods used to test hypotheses about those parameters. Still considering the example given above, suppose we have a special interest in whether or not the mean is equal to one. We then choose as our <u>null hypothesis</u>, the hypothesis to be tested, the equation $\mu_x = 1$. We saw above that 1.0 lies outside the 99% confidence interval on the mean. Hence it is unlikely that we would observe what we did in fact observe if the mean were one. We reject the null hypothesis at the 1% level. If the null hypothesis were that $\mu_x = 2.9$, we would reject it at the 5% level. Consider the general null hypothesis $\mu_x = Q$. When the variance of x is known, the following statistic provides a test of the null hypothesis:

$$(\hat{\mu}_x - Q)/(\sigma_x/\sqrt{T}).$$

If this statistic is greater than two in absolute value, the null hypothesis is said to be rejected at the 5% level. If it is greater than 2.6 in absolute value, the null hypothesis is rejected at the 1% level.

The relation of this statistic to the formula given above for the confidence interval should be clear. Note that you can never formally accept a null hypothesis: you can only fail to reject it.

In any statistical analysis, one takes as the null hypothesis some hypothesis of particular interest. In a study of the effect of fertilizer on crop yields, the logical null hypothesis is that there is no effect. In demand analysis, one sensible null hypothesis is that the price elasticity of demand equals minus one. A very common null hypothesis in connection with models of the form of (2.2) is that one of the true b's is zero. If b_j is zero, this means that, given the assumed functions, X_j does not affect Y. The variable X_j then does not belong in the equation. The hypothesis that b_j equals zero is a natural one to test, then, if we are not absolutely certain that X_j influences Y. If this null hypothesis is rejected at the 5% level, we say that the coefficient b_j is significantly different from zero (or significant) at the 5% level. In some sense, rejection of $b_j=0$ means that X_j probably does matter and should be left in the equation. If the null hypothesis cannot be rejected, the variable should generally be dropped and the equation re-estimated without it. This should be done because the fewer the coefficients being estimated from a given number of observations, the more reliable will be the estimate of each coefficient.

How do we test the hypothesis $b_j = 0$? We mentioned above that the least squares estimates of the b's are random variables, since they depend on the (unobservable) values taken on by the error term. If the error term is normally distributed, the b's will be also. The true means of these random variables are, as you might expect, the true b's - provided some assumptions discussed below are valid. The standard deviations of these estimates, however, are not known a priori. It is possible to estimate them, though, and all standard regression programs do this. The test outlined above is approximately correct when the standard deviations are estimated correctly. We will discuss in the latter part of the next section how the test must be modified to take account of the fact that the standard deviations must be estimated. Let us first examine the assumptions and procedures of multiple linear regression.

Regression: Method

For simplicity, we will write our basic model (2.2), in the following form:

$$(2.8) \qquad Y = b_1X_1 + \ldots + b_nX_n + e$$

The variables Y and $X_1, \ldots, X_n$ in this equation are, in general, functions of the variables appearing explicitly in equation (2.2). Defining new variables in this way costs us no generality and simplifies the notation.

We have spent a long time talking around least squares estimates, without defining just what such estimates are. We will now provide such a defintion. Suppose that we have any set of estimates of the b's in (2.8). As usual, call these estimates

$\hat{b}_1, \ldots \hat{b}_n$. We can then write, for each observation,

$$\hat{e}(t) = Y(t) - \hat{b}_1 X_1(t) - \ldots - \hat{b}_n X_n(t).$$

This equation defines the estimated error $\hat{e}(t)$, often called the residual, for each observation for the given estimates of the b's. Least squares estimation consists in choosing the b's so as to minimize the sum of squared residuals, $\sum_{i=1}^{T} \hat{e}(t)^2$. The references present the algebra involved in computing least squares estimates, but for our purposes a knowledge of these mechanics is not at all essential.

Are the estimates so obtained good ones? It all depends on how one defines "good", as there are a number of properties that are desirable in an estimation method. We will consider several of the most common and indicate when ordinary least squares possesses them.

First, an estimation method should be moderately simple, in the sense that the required computations are not unduly elaborate. Least squares is a member of the class of linear estimators, since it can be shown that the b's are linear functions of the observations on Y. Linear regression is thus a fairly simple estimation method.

A second desirable property is that the estimates not differ systematically from the true values. One measure of this property is called bias; an estimator of equation (2.8) is said to be unbiased if $E[\hat{b}_i] = b_i$ for all i. Least squares estimates will be unbiased in general if the following assumptions hold:

A.1 $E[e(t)] = 0$, all t. The error term is assumed to to have a mean of zero.

A.2 cov $[X_i(t), e(t)] = 0$, all i and t. The error term is not correlated with the independent variables.

A.3 It is never the case that, for $i \neq j$, $X_i(t) = a+bX_j(t)$ for all t. No two independent variables are perfectly correlated (move exactly together).

In time-series analysis, it is often the case that one of the X's in (2.8) is equal to the value that Y assumed in an earlier period. Such quantities are referred to as lagged dependent variables. When an equation contains one or more lagged dependent variables, the following assumption must be added to A.1 - A.3 in order to ensure that the estimates will be unbiased for large samples:

A.4 cov $[e(t), e(t+a)] = 0$, all t and a. The errors corresponding to different observations are uncorrelated.

(Whenever lagged dependent variables are present in a regression there will be a bias in small samples. This bias is usually neglected in practice as long as A.4 holds.)

A third property that a good estimator should have is efficiency. We would like the estimated values to be as close as possible on average to the true values. If we add one more assumption to the set A.1 - A.4, least squares estimates will be efficient in the sense that $E[(\hat{b}_i - b_i)^2]$ will be lower for all i than for $\hat{b}_i$'s generated by any other linear unbiased estimator. That assumption is

A.5 $E[e(t)^2] = \sigma_e^2$, a constant for all t. The variance of the error term is the same for all observations.

One last assumption is often made when linear regression is used:

A.6 e(t) is normally distributed for all t.

If assumption A.6 holds in addition to A.1 - A.5, least squares estimates are maximum-likelihood. This means, roughly, that the least squares estimates of the b_i maximize the probability of obtaining the data actually observed. Put slightly differently, the least squares estimates are those b's that make it most likely that we would observe what we actually did observe. This does not sound too powerful, but it is a very desirable property for estimators to have. Second, as we have mentioned, the estimates will be normally distributed. If A.6 holds approximately, both these properties also hold approximately. The mean of the $\hat{b}$'s will be the true b's, but the variances (and covariances) are of course not known a priori. Mention of the normal distribution suggests (again) that we can modify the hypothesis test we developed in the last part to test hypotheses on the individual coefficients, and we shall indeed do this shortly.

First, though, we shall discuss what happens to least squares when assumptions A.1 - A.5 fail. Assumption A.1, that the error term have zero mean, is not an issue if the regression equation contains a constant term. It will then be impossibel to tell whether the estimate of the constant corresponds to the true value of the constant term or to a non-zero mean value of the errors. But is is difficult to imagine a situation in which this distinction matters. A very important fact of least squares is that when there is a constant term (intercept) present in the estimated equation, the computed residuals sum to zero identically.

Violation of assumption A.2 is a problem in simultaneous equation models, as we discussed earlier. Also, this assumption is violated when lagged dependent variables are used as independent variables. In either case violation of A.2 will lead to biased estimates. Corrective action can be taken: see the references. This assumption cannot be tested using the data. It is a property of least squares estimates that the residuals will have zero (sample) covariance with each of the independent variables in the equation.

If assumption A.3 as stated above is violated, it is literally impossible to compute least squares estimates. (The computer will try to divide by zero - and foil). This is not a common problem; what often happens is that two or more of the independent variables are almost perfectly correlated. This problem is called multicollinearity. When two variables, say X_1 and X_2, move very closely together, it is hard for you or the computer to tell which changes in Y are due to changes in X_1 and which are caused by changes in X_2. If the two variables are expected to move together in the future, there is no problem in terms of forecasting. One variable is simply dropped from the equation, and when the regression is re-estimated the other picks up the effects of both on Y. If you are really interested in the separate impact of the correlated variables, though, the problem is far from simple. Occasionally, the theory underlying the equation may suggest a solution. The usually textbook anser is "get more data", but this may not be possible. Nature does not always perform the sorts of experiments we would like her to.

How do you tell if multicollinearity is causing trouble ? There is no hard and fast answer, from the nature of the problem. If all the coefficients are significant, there is usually nothing to worry about. If several coefficients are not significant, and dropping one or more variables makes some of these significant, multicollinearity is usally causing trouble. A glance at the correlation coefficients of pairs of independent variables (if available) may indicate the source of the problem.

Assumption A.4 is mainly an issue in time-series studies. When the error terms from different (usually adjacent) time periods are correlated, the problem is said to be one of <u>autocorrelation</u> or <u>serial correlation</u> of the errors. When this is present, least squares estimates are not efficient, and they are biased if lagged dependent variables are present. There are tests, involving the computed residuals, for simple sorts of autocorrelation,and corrections can be made to produce unbiased, efficient estimators. We will examine the most common test for autocorrelation in the last section of this chapter.

Besides indicated that the error-genrating mechanism is not what we would prefer, the presence of autocorrelation in the residuals may signal another problem. In fact, this other problem is usually the source of the observed serial correlation. Systematic movement of the estimated errors is usually caused by mis-specification. There are two types of mis-specification. First the functional form of the estimated equation may be incorrect. If the data came from $Y = b_1 + b_2 X + b_3 X^2 + e$ and the equation $Y = b_1 + b_2 X + e$ is estimated, there will be systematic (autocorrelated) errors because of the failure of the estimated equation to take account of the real non-linearity present. So one approach to the problem of serial correlation is to try different equations. Another kind of mis-specification is leaving out important variables. If the equation produces systematic residuals (differences between forecast Y and actual Y), it may mean that some variable that (1) influences Y and (2) moves systematically over time has been left out of the estimated equation. If alternative sensible functional forms don't get rid of the serial correlation, one might reconsider the basic independent variables being employed. Has something important been left out? If the answer is yes, the solution is to re-estimate with the omitted variable or variables present.

In an attempt to explain and solve a problem of serial correlation, it is sometimes useful to look at a graph of the estimated residuals gainst time andsee if anything significant suggests itself. This is nto generally productive, though, unless detailed knowledge of the sample period that is not embodied in the data is available. (Knowing in which years major wars took place is an example of information of this sort). A graph of serially-correlated residuals against time usually look like a sine wave plus noise, and is not especially informative.

Finally, if only assumption A.5 fails, least squares estimates are still unbiased. They are not efficient, though. If you an get some idea of how the variance at the errors varies from observation to observation, it is possible to transform the variable so that efficient estimates of the b's are obtained. The null hypothesis of equal variance, or homoscedasticity , can be tested against the alternative of unequal variances, heteroscedasticity. The references contain descriptions of the tests involved.

Before examining the output produced by a multiple regression program, there is one final issue we must deal with. We must consider testing hypotheses on the true coefficients in (2.8). Under assumption A.6, least squares estimates of these coefficients are normally distributed with unknown variance. Suppose the standard deviation of $\hat{b}_i$ is estimated to be $\hat{\sigma}_i$ by the regression program. (See the references for the algebra and statistics involved in constructing such estimates,) If the null hypothesis $b_i = Q$ is to be tested, the discussion in the last part of this section suggests that the following statistic be employed:

$$(2.9) \qquad t = (\hat{b}_i - Q)/\hat{\sigma}_i.$$

This is the so-called <u>t-statistic</u>. If the null hypothesis is true, it is distributed as Student's t with mean zero. (Recall that if the standard deviation of b_i were know a priori, this statistic would be normally distributed under the null hypothesis with mean zero and standard deviation one. See the defintion of the normal distribution). If b_i is not equal to Q, the distribution of t will clearly <u>not</u> have a mean of zero.

If the t-statistic is greater than two in absolute value, can be reject the null hypothesis at the 5% level? Almost, but not quite. The correct procedure involves consideration of the <u>degrees of freedom</u> of the regression. If there are T observation and n coefficients, the regression is said to have (T-n) degrees of freedom. The larger the number of degrees of freedom, intuitively, the more confident we are of our estimate of σ_i, and the more nearly we can act as if this quantity is know for certain.

In fact, unless T is at least equal to n, it is algebraically impossible to compute least squares estimates. (As before, the computer will attempt to divide by zero and fail). If T is equal to n, the estimated residuals will all be zero, and the $\hat{\sigma}_i$ will be zero also. (It is always possible to pass a straight line exactly through any two points in a plane, a plane can always be passed exactly through any three points in three dimensions, and so on). It is hard to have any realconfidence that the next observation encountered will be exaxtly explained by such an equation, and it is impossible to believe that simply because additional observations are not available we have found exactly the true coefficient. The larger is (T-n), the more information available per coefficient.

The exact density function of the t-statistic therefore depends on the degrees of freedom of the regression, on the "quality of the estimated coefficients and of their estimated standard deviations. For more than 20 degrees of freedom, t's larger than 2.09 in absolute value serve toreject the null hypothesis being tested at the 5% level. A t-valeu of 2.85 signals rejectionat the 1% level. If the regression involves fewer than 20 degrees of freedom, you must consult the tables in the references for the exact values of t that are needed to reject the null hypothesis.

Regression: Results

Suppose we have gathered or been given a body of data, we have verified the applicability of least squares, and we have gone to the computer and estimated the coefficients of a plausible regression equation of the form of (2.2) that incorporates all the information we can distill from economic theory and common sense. We now want to look at the statistics the computer will provide and we want to see what these numbers tell us about our trial equation.

The following are the main outputs of most regression programs:

1. Estimated Coefficients
2. Standard Errors of the Coefficients
3. t - Statistics
4. R^2 Statistic
5. Standard Error of the Equation
6. Durbin-Watson Statistic.

We shall discuss each of these in turn.

In terms of the notation we have been using, the estimated coefficients are the $\hat{b}_i$'s and the standard errors of the coefficients are the $\hat{\sigma}_i$'s, the estimated standard deviations of the estimated coefficients. The t-statistics are simply the ratios of the coefficients to their estimated standard errors. A glance at (2.9) indicates that the null hypothesis that the coefficient is zero is implicitly being tested. If the absolute value of the t-statistic is large enough to reject this null hypothesis at, say, 5%, the coefficient is said to be significant at the 5% level. If the coefficient also has the expected sign and a reasonable magnitude, there is a strong presumption that the corresponding independent variable does affect the dependent variable.

Besides the standard errors, most regression programs, provide the estimated variance-covariance matrix of the coefficients. The $\hat{\sigma}_i$'s are simply the square roots of the estimated variances. The estimated covariances can be used to test hypotheses and establish confidence intervals on linear combinations of the b's. The fundamental result that makes this possible is the following. If U and V are normally distributed random variables with mean μ_u and μ_v, variances σ_u^2 and σ_v^2, and covariance cov (u,v), the linear combination aU +bV, with a and b any constants, is normally distributed with mean $[a\,\mu_u + b\,\mu_v]$ and variance $[a^2\sigma^2_u + b^2\sigma^2_v + 2ab\ \mathrm{cov}\ (u,v)]$.

If for example, it is desired to test the null hypothesis that $b_1=b_2$, the following t-statistic is computed:

$$(\hat{b}_1-\hat{b}_2)/\sqrt{\hat{\sigma}_1^2 + \hat{\sigma}_2^2 - 2\widehat{\mathrm{cov}}(\hat{b}_1,\hat{b}_2)}$$

All the quantities under the square root sign are taken from the estimated variance-covariance matrix of the coefficients. There are still (T-n) degrees of freedom associated with this statistic.

Since forecasts of Y from a model of the form (2.8) are simply linear combinations of the $\hat{b}_i$'s, an estimated variance of forecast can be computed following this sort of procedure, and confidence intervals on the actual value of the dependent variable can be calculated.

The size of such an interval will depend on how well the regression equation fits the given observations and on how good the coefficient estimates are. Again, the references contain more detailed discussions of these points.

The fourth item on the list above, the R^2 statistic, is a measure of how well the model explains the movements of the dependent variable. It is defined by the following equation:

$$R^2 = 1 - [\sum_{i=1}^{T} \hat{e}(t)^2 / \sum_{i=1}^{T} (Y(t) - \hat{\mu}_Y)^2].$$

From this defintion, it should be clear that the least squares can be thought of as a procedure for maximizing R^2. The value of R^2 will always be between zero and one if the regression contains a constant term. It may be less than one if no constant is present. The usual interpretation of R^2 is as the percentage of the variance of the dependent variable explained by the regression equation . (If you know about analysis of variance, this should sound familiar. Indeed, linear regression and anlysis of variance are two ways of looking at the same statistical procedure). The R^2 statistic is said to measure the goodness of fit of the regression. Remember, though, that econometrics is not concerned only with obtaining R^2's close to one. The individual coefficients must be significant and reasonable, and the form of the equation must be sensible.

The F-statistic, which is printed by many regression programs, can be used to test for the significance of the R^2. That is, it provides a test of the null hypothesis that all of the true b's are zero, that no relationship is present. On this hypothesis, the quantity $[R^2/(n-1)]/[(1-R^2)/(T-n)]$ is distributed as F with (n-1) and (T-n) degrees of freedom. Large values of F serve to reject the nill hypothesis. This is generally a rather weak test, especially in time-series analysis, since rather low (say .30) values of R^2 are usually shown to be significant by the F-test. IF it is desired to use this test, one must consult one of the tables in the references or elsewhere.

The standard error of the quations, or the standard error of estimate, is simply the estimated standard deviation of the error term. Any forecast from the equation has a margin of error at least equal to one standard error. To really evaluate the accuracy of forecasts, however, confidence intervals must be computed, as outlined above. A 95% confidence interval on a forecast will be at least four standard errors wide.

Notice that the standard error will depend on the exact dependent variable used. An equation with dependent variable Y will, in general, have a larger standard error than the same equation with log (Y) as the dependent quantity. This is simply because the variance of Y is usually greater than the variance of log (Y), since, for Y greater than one, Y is larger than it logarithm. The real significance of the standard error is how large it is relative to the variance of the actual dependent variable used, and this is measured by the R^2 statistic.

The Durbin-Watson Statistic (DW) is used to test the hypothesis that adjacent errors are uncorrelated. This statistic is essentially worthless if lagged values of Y have been employed as independent variables in the regression. Our discussion applies only to the case where no lagged dependent variables are present. The formal model that underlies the DW test considers the error term e(t) as having been generated by a process of the following sort:

$$e(t) = \rho e(t-1) + u(t),$$

where ρ is a constant between minus and plus one, and adjacent values of u(t) are not correlated. This is called a first-order markov process, and ρ is the first-order serial correlation coefficient.

If such a process is generating the e's, the expected value of the DW statistic is approximately $2(1-\rho)$; the range of the DW statistic is 0 - 4. Values of DW less than two indicate a ρ between zero and one. Looking at the equation above, this means that successive error terms will be close together. With a value of ρ of .90, for instance, one would expect a long run of positive errors followed by a long run of negative errors, and so on. The pattern of residuals against time would generally look (as we remarked above) like a noisy sine wave. If ρ is between zero and minus one, on the other hand, positive errors tend to be followed immediately by negative errors, and vice versa. The first case ($0 \leq \rho \leq +1$) is called positive serial correlation, and the second situation ($-1 \leq \rho \leq 0$)

is referred to as negative serial coerrclation. Positive serial correlation is the more common the two problems.

What values of the Durbin-Watson statistic serve to reject the null hypothesis of no first-order autocorrelation, the hypothesis $\rho = 0$? Naturally, the critical values depend on the number of observations and the number of estimated coefficients - see the references for tables. The DW test has a rather bothersome particularity. The tables present values, labeled d_l, such that the hypothesis of no serail correlation is rejected if DW is less than d_l or greater than $(4-d_l)$. The tables also show another set of numbers, labeled d_u, such that the null hypothesis is retained if DW is between d_u and $(4-d_u)$. If DW is in the range of $[d_l-d_u]$ or $[(4-d_u)-(4-d_l)]$, nothing definite can be said. A safe course is to reject the null hypothesis if DW is outside the range $[d_u-(4-d_u)]$. A rough rule of thumb is that a value of DW outside the range 1.5 - 2.5 signals trouble.

References

We have come to the end of our informal treatment of multiple regression. We have (hopefully) made no really incorrect statements, but neither have we verified most of what we've said. Listed below are some textbooks that can be read by students with no mathematics beyond elementary calculues which examine econometrics more rigorously than we have been able to do here. Many of the difficult points that we have glossed over are dealt with in these texts, useful tabeles are presented, and techniques beyond ordinary least squares regression are developed. It is highly recommended that the student look at one or more of these books before attempting any estimation. The texts are arranged roughly in order of increasing difficulty:

1. M. J. Brennan, Preface to Econometrics, (Cincinnati: South-Western, Second Editinn, 1965).

2. E. J. Kane, Economic Statistics and Econometrics, (New York: Harper and Row, 1968).

3. L. R. Klein, An Introduction to Econometrics (Englewood Cliffs, N.J.: Prentice Hall, 1962).

4. C.E.V. Leser, Econometric Techniques and Problems, (New York: Hafner, 1966).

5. J. Johnston, Econometric Methods, (New York:McGraw-Hill 1963).

Kane's text has a rather complete set of tables. The other references are less useful in this regard. The various editions of the CRC handbook of Standard Mathematical Tables contain tables for the F and t tests. The books by Leser and Johnston assume a basic knowledge of statistics.

CHAPTER III

The Estimation Exercises

The nature of these exercises may be clarified a bit by the presentation of the sort of role-playing you might engage in. Imagine that you're on the headquarters staff of a large congolmerate, or of the Council of Economic Advisors - or even of Nader's Raiders. Your task is to investigate first the demand structure and then the cost functions for a particular industry. Your diligent assistant has come up with some historical information, which is described in the next section. Your task is to intelligently use these data and to present your conclusions in an informative report. The main difference between these exercises and real-world problems of the sort described is that the data here are better and the situation investigated is simpler than is usually the case in reality.

After discussing the nature of the industry and of your data, we turn to a rather brief presentation of commonly-made mistakes in applied econometric work. This section could be much longer; possible errors are legion. But experience suggests that the only way to really gain an understanding of the use of linear regresion is to actually use it. Scientific though economics tries to be, applied economietrics is still somewhat of an art. The last section of this chapter discusses the essential elements that should be present in good staff reports on these two exercises.

The Industry

All cost data refer to the operations of a typical firm in the industry under study. Similarly, the demand data refer to the demand experienced by an average firm in periods in which all firms were selling at the same price.

The population and the genral price level (cost of living) have not changed in the period for which data are provided. Aggregate disposable income (consumer's income after taxes) has changed, however, and you will be given the values it attained.

In the period for which data are available, the industry had a large number of firms, the number of firms was the same in all years. (You will be given the number of forms in the forecasting problem; you do not need this fact here). Each firm operated one plant producing a non-durable consumer good. All plants were and are identical, and all have a capacity of 100,000 units of output per year. No more can be produced under any circumstances. There has been no technical change in this industry; the same production process was in use in all years for which data is provided.

The variable factors of production were and are used in strictly fixed proportions. An index of the price of these factors will be provided along with other cost data if these prices have changed over the period. (If labor were the only variable factor, this index would be equal to the wage rate in each year divided by the wage rate in an arbitrary base year.) Cost figures include the opportunity cost of capital and entrepreneurship. Notice that the assumption of unchanged plants means that some factors or production were fixed throughout the period.

If this were all you knew about the industry, you would be faced with a rather severe problem. Price and quantity would be determined by the intersection of the industry supply and demand curves. Since disposable income and the prices of the variable factors of production varied over the period, both curves would be identified, as we discussed in the last chapter. But any sensible model of the industry would have to take account of the fact that price and output are determined by two simultaneous equations. Ordinary least squares could not be used for your investigations, as it would produce biased estimates, estimates that differed systematically from the true coefficients.

One fact about the industry, stated in Chapter I, makes possible the application of ordinary linear regression. At no time in the past has the market price of this commodity been determined wholly by supply and demand. The output of each firm in the industry was either determined directly by the government or fixed by arbitrary non-price rationing of an essential input factor. Thus firms' costs did not affect supply or price. A simultaneous equation model is not appropriate; use of least squares is.

Estimating Economic Relationships

A first warning. When you are given a listing of the data, there is a natural tendency to pore over it in an attempt to discover significant relationships. This is generally harmless of not too productive. But many students waste hours trying to obtain simple insights from a large number of figures. The computer can sumarize and analyze data better than you can with a pencil and paper. Use the computer for what it can do best.

But this remark calls for a second caveat. The computer cannot think, and it knows no economic theory. Since the number of equations you can estimate will be limited, you must do some analysis before going to the computer. The demand function is a model of consumer decision-making, and the cost function reflects both technology and managerial choices. You must make sure that the equations you estimate reflect these characteristics of the underlying relationships. Are the right variables present? Are the functional forms economically sensible? Are they needlessly complex?

Do not try to use the computer to select "important" variables before estimating cost and demand functions. When one variable depends on two or more other variables, regressing the dependent variable on the independent variables on-at-a-time can produce highly misleading results. To dramatize this point, consider the data presented in Table III1; Examine first only the variables Z and X. Notice that exactly half of the changes in X are associated with changes in Z in the opposite direction. No relation is apparent between X and Z. If only Y and Z are considered, exactly the same conclusion is reached. Yet an examination of all three columns reveals that Y is always exactly equal to the sum of X and Z. All relevant variables must be present in a regression equation if one is to make sensible judgements about the importance of any of them.

Another commom mistake is to estimate identities or near-identities. Total revenue does depend on price and output; it is identically equal to their product. Any equation with revenue as a dependent variable and price and quantity as independnet variables will produce a good fit. If the logarithms of these quantities are used, the fit will be perfect. But the equation will have no meaning at all in either case. If the same variable is present on both sides of an estimated equation, either directly or implitcitly via an identity, be certain that there is a good, theoretically sound reason for it to be there. It may be appropriate to express all quantities in a demand equation in per capita terms when population is changing, but it will nev er make sense to multiply all variables in a demand relation by the selling price of the good in question.

After you have estimated a few sensible quations in either problem, it is again time for thought. Are any of your origincal hypotheses concerning important variables refuted by the results of estimation? Are any new hypotheses suggested? (Recall the use of the DW statistic as a signal of mis-specifications). There is a great temptation to pore over residuals to find a functional form that fits the data slightly better than the ones first employed. There is nothing fundamentally wrong with this, except that it violates the

Table III.1

Some Illustrative Data

Year	Y	Change in Y	X	Change in X	Z	Change in Z
1	10		1		9	
2	15		7		8	
3	13	-	8	+	5	
4	14		7		7	
5	12		3		9	
6	13		6		7	
7	14		4		10	

the venerable principle of Occam's Razor. Whenever a simpler equation performs almost as well as a more complicated one, the simpler form should be retained. This means that variables whose coefficients are not significant would generally be dropped from the equation in order to increase the degrees of freedom available. Bending over backwards to get a slightly better fit (R^2) offends most instructors. A final reason for considering only simple functional forms is that you will be required to use your estimated relationships in calculations later. The more complex the function you choose, the more laborious will be your computations.

Your Reports

In writing up your investigations of the cost and demand structure, it is useful torecall the scenario sketched out at the start of the chapter. What would a good report look like in that sort of situation? The answer is clear; your supervisor would want to know what you did, why you did it, and what conclusions your research led to.

Keeping this in mind, a good report should first discuss what economic theory has to say aboutthe form of the (cost or demand) relation being estimated. On the basis of this reasoning, you should present and defend the equations you estimated. The heart of the paper should consist of your selection and defense of your best estimate of the relation under study. Both economic and statistical properties should be considered.

Finally, the nature of the demand or cost structure you

estimated should be discussed. The relevant demand elasticities should be computed, and their implications examined qualtitatively. (Since these elasticities are not usually constant, use the average values of the dependent and independent variables in your computations.) Note that the cost data are short-run figures (no change in plant), so nothing can be said baout returns to scale in production. The value and behavior of such things as marginal cost, average variable cost and average fixed cost should be discussed, however. In qualtitative terms, what does the cost function look like?

Empirical research is always judged on the basis of the quality of the economic and statistical analysis. Goodness of fit is a desirable property in any estimated relationship, but it will not determine grades for this exercise.

CHAPTER IV

Competitive and Monopoly Pricing

In this chapter, we shall derive and discuss the conditions for competitive and monpoly equilibrium in the usual timeless model. The reason these market structures can be handled analytically is that there is a determinate relation between the industry demand curve and the demand curve facing each individual firm. Un-der monpoly, the industry demand curve is the demand curve facing the single firm present. If the market is competitive, each firm takes the going price as given beyond its control, though of course all firms together can sell only as much as industry demand permits. Under oligopoly, when there are few selers, the demand curve facing each firm depends on the expectations the firm has about its competitors' actions and reactions. In such a situation, few neat mathematical results are possible. We shall deal qualtitatively with oligopoly in Chapter VI of this manual.

Three basic elements will underly our analysis in this section. First is the industry demand curve, giving the number of units of its output that the industry can sell at each and every price. Whether the industry is competitive or monpolized, this function will give the relation between price and total industry sales.

The second element is the set of cost functions of the plants in the industry. We shall assume that each plant chooses the factors of production it uses to produce any output level so as to minimize the total cost of production. Thus he cost curves we shall analyse give the minimum cost of producing any output (less than capacity) in a given plant. These curves depend on the technology of production and conditions of factor supply. Throughout, we shall assume that the opportunity cost of the capital embodied in each plant is included in that plant's fixed cost.

Finally, we assume that each firm, be it monopolist or pure competitor, seeks to maximize it sprofits. It is this assumption that gives us definite results, for it makes it possible for us to use the differential calculus to obtain exact and (fiarly) simply solutions.

The next section will discuss the mathematical techniques employed in the remainder of the chapter. References listed at the end of the chapter go into these methods in more detail; our purpose here is simply to introduce the basic tools and concepts. We then consider short-run equilibrium in monopoly and competitive situations. In the short run, by assumption, the number and size of plants in the industry is fixed. The last part of this section analyzes long-run competitive and monpoly equilibria. That is, we consider equilibrium conditions when the number and size of plants can be changed. Throughout, we assume a constant technology and perfectly elastic supplies of factors of production.

Basic Mathematical Techniques

Suppose some variable y is a function of another quantity x, according to $y = f(x)$. It is desired to choose x so as to maximize y. If $x=x^*$ maximizes y, it will be the case that $df/dx = 0$, where the derivative is evaluated at $x=x^*$. That is, the vanishing of the first derivative is a necessary condition for a maximum; any point that is a maximum satisfies it.

This is not a sufficient condition, though. We cannot say that any point that satisfies $df/dx = 0$ maximizes y. If f(x) is convex (looks like a soup bowl right side up), the first derivative will be zero at the minimum of the function. The function f(x) must be concave (sahped like an upside-down soup bowl) when the first derivative vanishes for a maximum to be indicated. Also, there may be several points at which $df/dx = 0$ and the function is concave. Each of these corresponds to a local maximum, a point where y is larger than at all nearby points. The largest of these local maxima is the global maximum.

In the remainder of this discussion, we shall present necessary conditions for optimality in some economically interesting situations. We shall assume that the functios involved are shapred so that (1) these conditions do singal a maximum and (2) there is a unique point satisfying these conditions. That is, we assume that the sufficient conditions for a global maximum are satisfied. Sufficient conditions for the problems we shall discuss are presented in the references listed at the end of this Chapter.

The basic problem in what is usually called marginal analysis is of the following sort. Suppose that a decision-maker has two activities under this control. Let x and y be the levels at which these activities are to be carried out. We denote the benefits from these activities as B(x,y), and the costs as C(x,y). The decision-maker seeks to maximize (B-C). The necessary conditions for a maximum, found by setting the first partial derivatives of (B-C) equal to zero, are

$$(4.1) \qquad \frac{\partial B}{\partial x} = \frac{\partial C}{\partial x} \quad \text{and}$$

$$\frac{\partial B}{\partial y} = \frac{\partial C}{\partial y}$$

That is, the marginal benefits from each activity must equal the marginal costs of that activity.

In the case of a firm, x and y could be the outputs of two products. Then benefits are total sales revenue, and costs are production costs. For the Department of Defense, x and y may be the numbers of two kinds of nuclear delivery systems. Costs are then procurement and operating costs, and benefits are number of targets destroyed (on average). It should be clear that if (4.1) signals a negative value for x and y, that activity should not be undertaken.

A simple example will illustrate the use of conditions (4.1). Suppose $B(x,y) = 4x + 6y$, while $C(x,y) = x^2 + y^2$. Setting the partial derivatives of (B-C), with respect to x and y equal to zero we obtain

$$4 - 2x = 0 \quad \text{and}$$

$$6 - 2y = 0$$

Solving, we have x=2 and y = 3. Consideration of the shape of the function (B-C) should convince you that it is indeed maximized at this point.

An interesting special case of this problem is the following. Suppose there is only one output of interest, x. Benefits are a function of the amount of x produced. Suppose further that x can be produced two ways. The first way has a cost function $C^1(\cdot)$, and the second has cost function $C^2(\cdot)$. Let x_1 be the amount of x produced the first way and let x_2 be the amount produced the second way, so $x = x_1 + x_2$. Setting the first derivatives of $B-C^1-C^2$ equal to zero, we obtain the necessary conditions.

$$(4.2) \quad \frac{\partial B}{\partial x} = \frac{\partial C^1}{\partial x_1}, \text{ and}$$

$$\frac{\partial B}{\partial x} = \frac{\partial C^2}{\partial x_2}$$

These equations imply that the marginal cost of producing the activity must the same for all techniques employed. This means for instance, that a firm operating several plants should operate them so that the marginal cost of production is equal for all plants actually used. This is an efficiancy condition for production, in the sense that no matter how much x is produced, if it is produced at minimum cost the marginal costs of all plants or methods employed must be equal (If some methods or plants have marginal costs that are too high for this condition to be met at a given output level, those methods or plants should obviously not be used.)

So far, we have not explicitly considered any constants on the levels of the activities of interest, though we have mentioned that usually the activities must be carried on at non-negative levels. Suppose now that we have two activities x and y, and that these must be carried on so that the constraint equation $f(x,y) = k$ is satisfied, where k is some constant. This constant could represent, for instance, the amount of money a household has available for the purchase of two commodities, x and y. The constraint equation would then take the form $P_x x + P_y y = k$, where the P's are the prices of x and y.

In order to maximize (B-C) subject to the constraint equation $f(x,y) = k$, we can employ a useful technique developed by Lagrange. Write down the Lagrangian expression

$$L = B(x,y) - C(x,y) + \lambda[k - f(x,y)].$$

Notice that the quantity in brackets must be zero. Necessary conditions for a maximum are that the first derivatives of L with respect to x, y, and the <u>Lagrange multiplier</u>, λ , be zero. Formally, these conditions are

$$(4.3) \quad \frac{\partial B}{\partial x} - \frac{\partial C}{\partial x} - \lambda \frac{\partial f}{\partial x} = 0,$$

$$\frac{\partial B}{\partial y} - \frac{\partial C}{\partial y} - \lambda \frac{\partial f}{\partial y} = 0 \text{ and}$$

$$k - f(x,y) = 0.$$

These three quations serve to determine the three unknowns d, y, and λ It can be shown that, if the functions involved are such that (4.3) does signal a maximum, λ is the derivative of (B-C) with respect to k. That is, λ measures the marginal value of increasing k. In many economic applications, k is the amount of some fixed resource available to the decision-maker, and λ then has the interpretation of the marginal value of that resource. If there are two such resources, there will be two constraints, two multipliers, and so on.

Let us illustrate the use of conditions (4.3) by a slight extention of our earlier example. We wish to maximize

$$B(x,y) - C(x,y) = 4x + 6y - x^2 - y^2$$

but now we assume that x and y must satisfy the constraint equation

$$x + y = k.$$

Here the Lagrangian is

$$L = 4x + 6y - x^2 - y^2 + \lambda [k - x - y],$$

Differentiating, we find the three equations corresponding to conditions (4.3) to be

$$4 - 2x - \lambda = 0$$

$$6 - 2y - \lambda = 0$$

$$k - x - y = 0$$

For k=1, the solution to this system of three equations in three unknowns is easily seen to be $x = 0$, $y = 1$, and $\lambda = 4$. If the constant "k" were the available input of some resource costing $3.50 a units, and if the maximand were dollar profit, the value of λ indicates that a small additional purchase of the resource in question would be worthwhile. For $k = 2$, the solution to the three equations above is $x = 1/2$, $y = 3/2$ and $\lambda = 3$. At this point, additional purchases at a price of $3.50 would not be profitable.

Suppose that k were equal to ten. We showed above that the unconstrained maximum of (B-C) occurred at x=2, y=3. If k represents the available quantity of a scarce resource, the decision-maker will never use more than five units unless he is forced to. This leads to consideration of a more general problem, where the constraints take the form of inequalities rather than equalities.

Consider maximizing (B-C) subject to the constraint inequality $f(x,y) \leq k$. If the constraint involves a scarce resource, this inequality says that the decision-maker may employ up to k units of the resource, but he need not use all k units available. For example, k might be the maximum possible output of a plant. Problems of this sort may be solved by using the Kuhn-Tucker Theorem. To employ this technique we proceed as above to form the Lagrangian. Then differentiate, as before, to obtain the first two of conditions (4.3). Those two equations are necessary conditions for this more general problem. The Kuhn-Tucker Theorem then says that one and only one of the following coditions must hold.

(a) $f(x,y) < k$, and $\lambda = 0$

(b) $f(x,y) = k$ and $\lambda > 0$

In case (a), the constraint is said to be not binding, and the marginal value of the fixed resource in question is zero. It would not pay the decision-maker to purchase additional units of the resource, since he maximizes (B-C) without using the full amount already available. In case (b), the constraint is said to be <u>binding</u>. Additional units of the fixed resource do have positive value. (Whether this value is greater than the cost of the resource is, of course, another question.) In any particular problem, the solution will be either of form (a) or form (b). It is possible in theory to have $f(x,y) = k$ and $\lambda = 0$, but this almost never happens in practice. In the numerical example discussed above, if the constraint is $x + y = k$, case(a) can easily be seen to apply whenever k is greater than five. Only for $k = 5$ do we have $\lambda = 0$ and $x+y = k$.

In simple situations trial and error may be employed to find the solution. Assume that case (a) obtains and solve for x and y with $\lambda = 0$. If the constraint is satisfied, the problem is solved. If not, assume that case (b) holds and find a solution in which $f(x,y) = k$. With one or two constraints, the trial and eror process is not difficult.

When there are N constraints, there are 2^N possible cases to be considered, and when N is larger than two or three, the computation involved in trial and error becomes altogether too involved. In such cases, the techniques of mathematical programming are employed. These methods make possible the solution of complex problems of this sort without resort to trial and error. The simplex method for linear programming is the best-known procedure; it is used when the functions B and C and all the constraint inequalities are linear function of the activity levels.

This short section has presented a lot of material superficially. For our purposes here and in the next chapter, this is hopefully adequate. The references at the end of the chapter cover these topics in more depth; the interested (or confused) student should examine them.

Short-run Equilibrium

All students of economics know - or should know - that a competitive firm sets price equal to marginal cost, while a monpolist sets marginal revenue equal to marginal cost. In the short run, if a competitive industry and a monpolist have the same plants available, the monopolist will produce less and earn larger profits than the competitive industry. The community is generally better off under competition. All this is cov ered in standard economics texts.

In this section, we shall apply the mathematical tools just presented to the problem of competitive and monopoly pricing. Our aim is to develop the methods needed to solve the forecasting problems of the next chapter. Accordingly, we shall make some simplifying assumptions. First, we assume that the prices of the factors of production used by the industry do not depend on the industry's output. This is the assumption of perfectly competitive factor markets. We suppose the industry considered to be only a small purchaser (relative to the market) fo the factors of production it uses.

Secondly, we assume that the industry is composed of N identical plants. With no changes in technology, this is a reasonable assumption. If there is a best sort of plant, all plants built will be of this type. If there is no best plant, all plants will be identical by definition. The term "plant" must be interpreted broadly in this context; an optimal "plant" may consist of one machine, one man, and one hundred square feet of floor space with a roof over it. Then a plant in the usual sense will consist of a number of these plants at the same location.

Let q be the output of a typical plant, and write the cost function of a typical plant as C(q). We denote the short-run marginal cost of production in the typical plant as MC(q). By definition, MC(q) = dC/dq. We assume that no plant can produce more than k units of output in any period. We shall refer to k as capacity output. (If there is no upper limit on output on the short run, k can be made very large so that the capacity constraint is not binding.)

We shall denote the industry's total demand by Q and the price at which these sales are made by P. These two quantities are connected by the industry demand relation P = P(Q). We shall first examine short-run competitive equilibrium under these assumptions. The remainder of this section considers the equilibrium of a monopolist.

As long as there are a large number of firms and the market is perfectly competitive, it does not really matter how many plants each firm owns. Each plant will be operated so as to maximize the profit it earns, on the assumption that the firm's activities will not alter the market price. Thus each plant in the industry will be operated so as to maximize

$$(4.4) \qquad P\,q - C(q),$$

where P is treated as a parameter beyond the control of the firm. The quantity (4.4) must be maximized subject to the constraint $q \leq k$.

Let us first solve the problem algebraically; we shall then examine the geometric interpretation of the solution. As in the last section, we form the relevant Lagrangian:

$$(4.5) \qquad L = P\,q - C(q) + \lambda\,[k - q].$$

From our discussion in the last section, the first necessary condition for maximum profits is found by differentiating the Lagrangian with respect to q, treating P as a constant. Using the Kuhn-Tucker Theorem, the set of necessary conditions is

$$P - MC(q) - \lambda = 0 \quad \text{and either}$$

$$(4.6) \qquad \text{(a)} \quad q = k \text{ and } \lambda > 0, \quad \text{or}$$

$$\text{(b)} \quad q < k \text{ and } \lambda = 0.$$

For a given value of P, we can solve these conditions for the output of a typical plant. Note that P will equal MC unless P Is greater than MC(k), marginal cost at capacity output. For prices greater than this value, all plants in a competitive industry will produce capacity output. The multiplier λ measures the amount by which profits would rise if k were increased. If q is less than k, it is clear that λ should be zero.

Now that we know how firms' supplies will vary with market price, it is a rather simple matter to determine industry equilibrium. Suppose that we have solved conditions (4.6) for q as a function of P. CAll this function $q^s(P)$; this is the short-run supply curve of a typical plant. Then $Nq^s(P)$ is the industry's short run supply schedule. Solving the demand function for Q, we can write it as $Q = Q^d(P)$. In equilibrium, supply equals demand, so we have

(4.7) $Nq^s(P) = Q^d(P)$.

This equation can be solved for P, since N is given. ONce P is known, Q and q(=Q/N) can be calculated from the demand schedule.

The steps in this solution are really quite simple. First we found out how much the typical prfit maximizing plant would supply at each price. We then multiplied by N to obtain the industry's supply as a function of price. Finally, the intersection of this function with the industry demand curve was used to determine equilibrium price and output.

We can further clarify the nature of the equilibrium by considering Figure IV.1. This graph refers to a single plant, where we define $\tilde{q}^d(P) = Q^d(P)/N$. This curve shows how much each plant would sell if all plants charged the same price. (Recall the demand function you estimated in the last chapter). It is not the demand curve facing each plant, since each plant is too small to in any way affect the price charged by its competitors. But q^d must equal

Figure IV.1

<u>Short-Run Competitive Equilibrium</u>

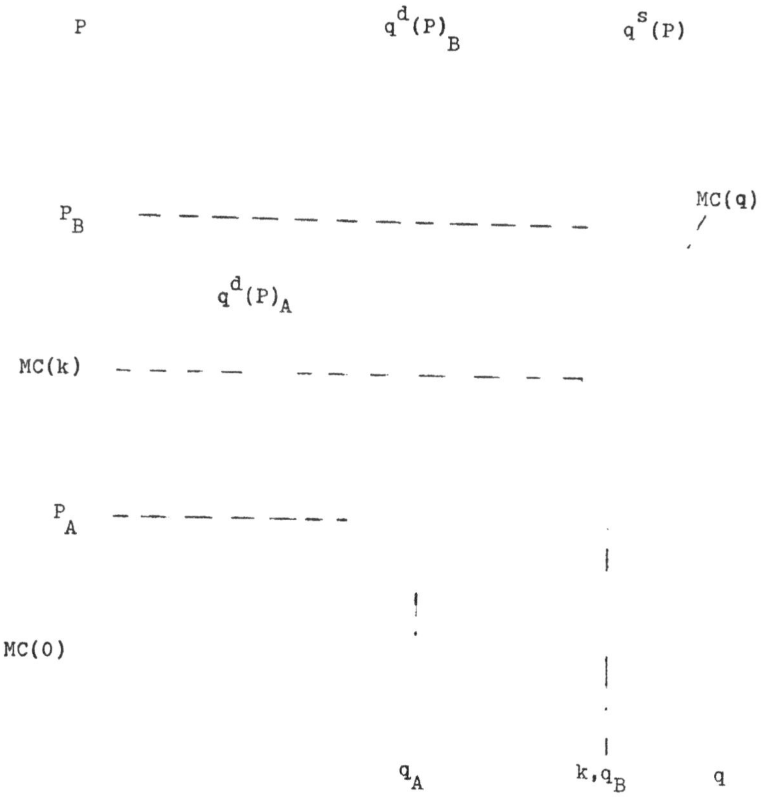

q^s in equilibrium; Supply must equal demand, and the market must clear.

Suppose that the industry demand function and the number of firms are such that $q^d(P)_A$ applies. At equilibrium, price is equal to marginal cost, and output is below capacity. Were demand to shift up to $q^d(P)_B$, price could not equal marginal cost. The two curves intersect only for an output above capacity. Clearly a rational firm will produce all it can, k, and the market price will settle at P_B. In the first of these cases, the multiplier will be zero, while with a larger demand and capacity output being produced, the multiplier will be positive. Since price is above marginal cost, another unit of capacity would be profitable for each firm individually, though the industry as a whole might lose profits if more capacity were purchased.

The method of analysis used when the industry is monopolized are the same as those employed above, with one large difference. The monopolist is correctly aware that his production decision will fix market price. He is still interested in maximizing the difference between revenue and cost, but he will treat matket price not as a parameter, but as a variable which he determines.

As immediately above, there are N identical plants in the industry. But now they are under the control of a single firm. Since the plants are identical, they will each produce the same amount under any rational plan. Thus $q = Q/N$ here as under competition. (If the plants were not identical, cost-minimizing production would require that all plants in operation have the same marginal cost; see the last section.)

Using the same notation as above the monopolist is trying to maximize

$$(4.8) \quad P(Q)\,Q - N\,C(Q/N),$$

subject to the constraint $Q/N \leq k$. The relevant Lagrangian is

$$(4.7) \quad L = P(Q)\,Q - N\,C(Q/N) + \lambda[k - Q/N].$$

Since N is fixed in the short run, the necessary conditions become simply:

$$[P + Q\,dP/dQ] - MC(Q/N) - \lambda/N = 0, \text{ and either}$$

$$(4.8) \quad \text{(a)} \quad Q/N = k \text{ and } \lambda > 0, \text{ or}$$

$$\text{(b)} \quad Q/N < k \text{ and } \lambda = 0.$$

The quantity in brackets in the first of these conditions is simply the derivative of total revenue with respect to output, the monpolist's marginal revenue. Since it varies along the demand curve, marginal revenue is clearly a function of output, Q.

To clarify our formulae, we define

$$MR\,(Q) = P(Q) + Q\,dP\,(Q)/dQ.$$

Then the first of conditions (4.8) becomes.

$$(4.9)\quad MR(Q) - MC(Q/N) - \lambda/N = 0.$$

Conditions (4.8) state that marginal revenue must equal marginal cost if plants are not producing capacity output. If capacity output is produced, it must be the case that marginal revenue is above marginal cost. Notice that if "marginal revenue" is replaced by "price" in the last two sentences, they become the conditions for a competitive equilibrium. But the difference is clear; marginal revenue will be less than price at all output levels, so price will always be above marginal cost for the monopolist. Since marginal cost is always positive, marginal revenue must also be positive at equilibrium. It is easy to show that this means that demand must be elastic at monopoly equilibrium. It shoauld also be noted that a monopolist's production decision depends not only on price, but also on the rate of change of price with respect to output, dP/dQ. Thus a monopolist has no supply curve; the quantity he will supply to the market depends on the whole demand curve, and it is uniquely determined by the shape and level of that curve.

In any particular problem, the path from conditions (4.8) to a solution is rather short. One first seeks a solution of (4.9) with $\lambda = 0$. IF this solution satisfies $Q/N < k$, plant and firm output have been found. IF not, it must be the case that $Q/N = k$. Once output has been determined, the industry's demand function will yield the price at which this output can be sold.

As we did in the competitive case, we now proceed to present the solution for a typical plant graphically. For a fixed number of plants, we can write marginal revenue as MR(Nq), where, as before, q is the output of a typical plant. We shall write the demand relation as $P = P(Nq)$. Solving this for q, we would obtain $q = q^d(P)$, the function we discussed in the competitive case.

Consider now Figure IV.2. The cost structure for a typical plant is represented by the curve giving short-run marginal cost as a function of output. If the demand structure is such that $P(Nq)_A$ applies, the solution is the usual textbook sort. Each plant will produce q_A, the firm will produce Nq_A, and price will be P_A. Marginal cost is equal to marginal revenue. Under demand regime B, marginal revenue is above marginal cost, and all plants are producing capacity output.

Long-Run Equilibrium

In the long run, both the kind and number of plants in an industry can change . Were we to simply add more variables and functions and mechanically proceed to derive necessary conditions, it is almost certain that the results would convey no insights at all. We must make some additional assumptionsin order to obtain useable results.

First we shall strengthen the assumption of identical plants. We now assume that any new plant built must be identical to the plants presently operating. Further, we assume that there are no economies or diseconomies of multi-plant operations. We can thus write the cost of efficiently producing an output Q in N plants as

Figure IV.2

Short-Run Monopoly Equilibrium

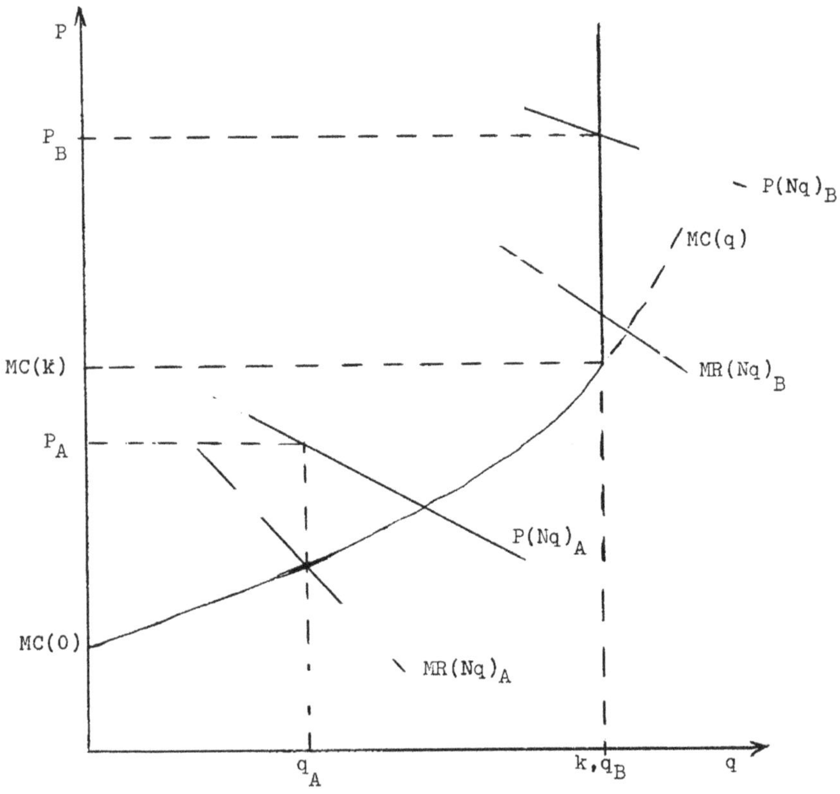

$N\ C\ (Q/N)$,

as in the last section. The function C(;) is the same one employed in the short-run case, but now N may vary.

The long-run equilibrium condition for a monopolist is the same as the short-run condition. The firm will maximize its profits. In the long run, though, the number of plants operated is under its control. Firms in a competitive industry will also try to maximize profits in the long run as in the short run. But we shall assume totally free entry into and exit from the competitive industry. This means that the profits earned by each plant, though they are being maximized, but be zero in long-run equilibrium. Postive profits will attract new firms; losses will force some firms to leave the industry. (As usual, by profits we mean earnings in excess of opportunity cost.)

The short-run equilibria derived in the last section were characterized by the relations between price and marginal cost and

between marginal revenue and marginal cost. There we were considering short-run marginal cost, the incremental cost of increasing output from a given plant. Here one might expect to encounter equations relating long-run marginal cost to price and to marginal revenue. Long-run marginal cost should be the incremental cost of adding to output when the possibility of varying N is taken into account. This discussion turns out to be substantially correct, as we shall see shortly. The algebra involved in the solutions will enable us to get a better idea of what long-run marginal cost is.

Since there is only one equilibrium condition (maximum profits) to worry about in the case of monopoly, we shall consider that situation first. We shall begin by treating the number of plants, N, as a continuous variable. Later we shall deal with the case where N must be an integer. Given the capacity constraint, the Lagrangian for this problem is:

$$(4.10) \quad L = P(Q)\,Q - N\,C(Q/N) + \lambda\,[k - Q/N].$$

Differentiating with respect to Q with respect to N, we have the necessary conditions for a maximum:

$$MR(Q) - MC(Q/N) - \lambda/N = 0,$$

$$(4.11) \quad - C(Q/N) + (Q/N)\,MC(Q/N) + (\lambda/N)(Q/N) = 0 \text{ and either}$$

(a) $Q/N < k$ and $\lambda = 0$, or

(b) $Q/N = k$ and $\lambda > 0$.

Consider case (a) first. Write $q = Q/N$. Then the second of conditions (4.11) becomes

$$(4.12)^{*} \quad C(q)/q = MC(q).$$

That is, marginal cost must equal average cost if all plants are operated below capacity. It is well-known (and easy to show) that when marginal cost equals average cost, average cost is at its minimum point. Thus, where the typical plant has minimum average cost occuring at a level of output below capacity., each plant will be operated at that level of output. This short of cost structure is illustrated by Figure IV.3(a). The minimum value of average total cost is labeled long-run marginal cost, LRMC, since it represents the total cost (including the cost of capacity) of producing an additional unit of output as cheaply as possible. In long-run equilibrium, each of the monopolist's plants will produce an output q*.

If average cost is above marginal cost at all output levels, and average cost is falling (the usual case), equation (4.12) will not be satisfied for any level of output below capacity. We then have case (b), depicted in Figure IV.3(b). All the monopolist's plants will be operated at full capacity in equilibrium .

We have used the second and third of conditions (4.11) to find th output per plant in long-run equilibrium. We now bring in the first condition to obtain the optimal number of plants - and thus total firm output. Solve the second condition for λ/N and substitute into the first. This yields:

$$MR(Q) = C(Q/N)/(Q/N).$$

Long-Run Marginal Cost

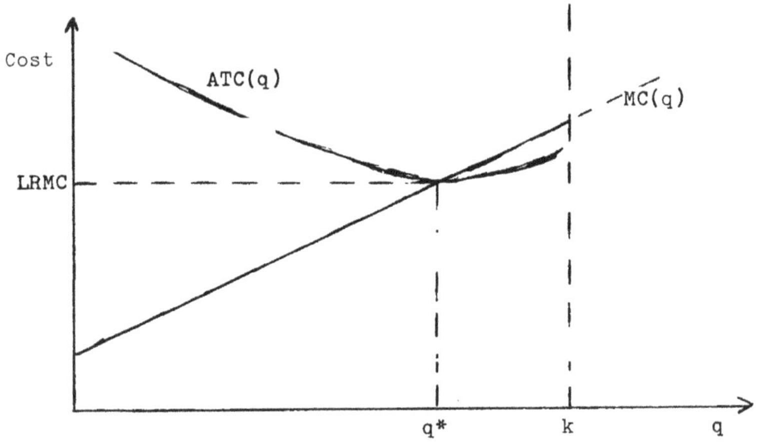

(a) minimum average total cost occurs at q* < k.

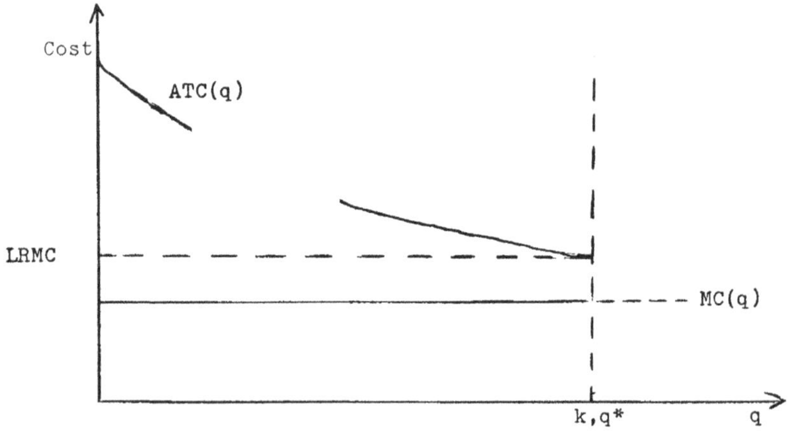

(b) minimum average total cost occurs at q* = k.

Since we know that each plant will be operated at the output level q^* that minimizes average total cost, this equation may be re-written as

$$MR(Nq^*) = C(q^*)/q^*,$$

which may be solved for N. Price is then obtained by substituting output, Nq^*, into the demand function, and the complete solution has been found.

This final step is illustrated by Figure IV.4 There Q_M is the monopolist's output, and P_M is his price. Division of P_M by q^* ill yield the optimal number of plants.

Notice that in btoh cases (a) and (b) all output has been produced by running each plant at its most efficient level. We divided the solution into two parts: finding that level and computing the optimal number of plants. If N is continuous, the long run marginal cost of the monopolist is constant and equal to the minimum value of average cost for the typical plant. Setting long-run marginal cost equal to marginal revenue yields the optimal quantity and the optimal number of plants.

When N must be an integer, one more step must be performed. Suppose the above analysis gives N as 79.34. Then, in almost all cases, the optimal number of plants will be either 79 or 80. The procedure

Figure IV.4

Long-Run Equilibria

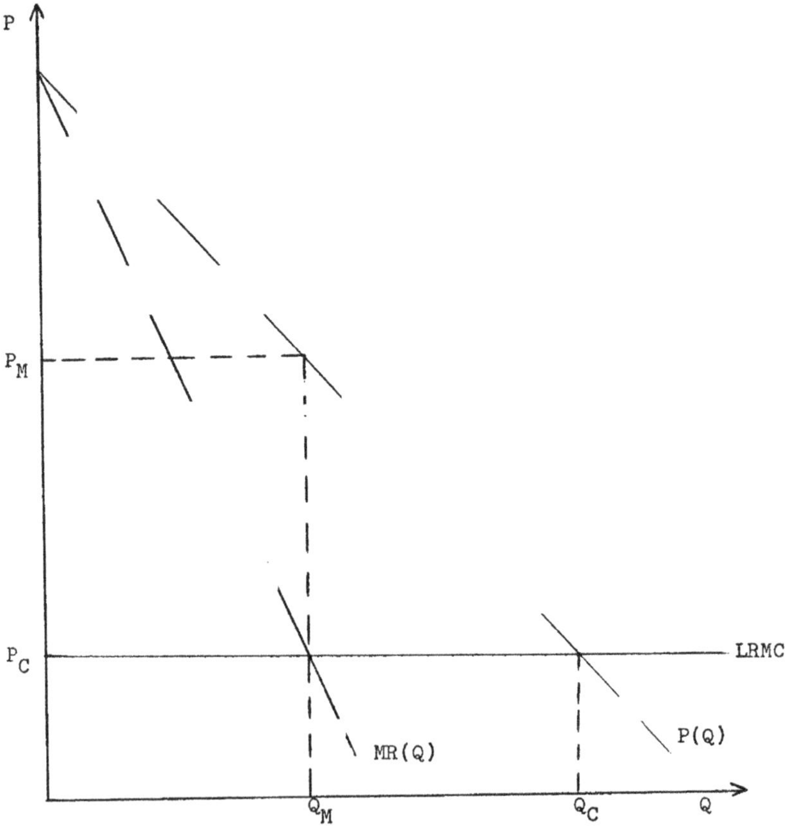

at this point is to assume N = 79 and solve the short-run problem, do the same for N = 80 , and choose that N for which profit is larger. Unless N is quite small, this method will yeidl the right answer.

Let us now turn to the competitive situation. It is simplest here to work with a single plant and to use $q = Q/N$ (= the output of the plant) as our main variable. The plant owner or manager will maximize his profits; his Lagrangian is

$$(4.13) \quad L = P\,q - C(q) + \lambda[k - q].$$

Since P is beyond the control of any single firm, the necessary conditions for a maximum are

$$P(Nq) - MC(q) - \lambda = 0 \quad \text{and either}$$

$$(4.14) \quad \text{(a)} \quad q < k \text{ and } \lambda = 0 \text{ or}$$

$$\text{(b)} \quad q = k \text{ and } \lambda > 0.$$

We have written P as a function of Q, since supply must equal demand in equilibrium.

Given either q or N, conditions (4.14) will determine the other quantity, We clearly need one more equation,and that is provided by the condition that in equilibrium profits (in excess of opportunity cost) must be zero. This condition is

$$(4.15) \quad P(Nq)\,q - C(q) = 0,$$

our additional equation.

We could simply solve these equations on the assumption that case (b) held. IF this assumption is verified, we have a solution. If not, case (a) must hold, and the resultant equations can then be mechanically solved. But there is a way of looking at these conditions that yields a bit more insight.

Suppose case (b) applies. Then we have

$$P\ MC(q) = C(q)/q.$$

That is, marginal cost equals average cost, and average cost must be, as we noted above, at its minimum point. If this equation does not hold for any q less than k, average cost must reach its minimum at capacity output, and case (b) will apply. What we have found is, perhaps not too surprisingly, jut what we found for the monopoly case: industry output will be produced by operating each plant in the industry at the point where average total cost is minimized.

The long-run supply curve of the industry is flat when N is continuous, and it is equal to the minimum value of average total cost. Industry output is determined by setting long-run supply equal to demand. As in the monpoly case, output per plant is determined only by cost conditions, and the number of plants is determined by demand conditions.

Thus Figure IV.3 applies to both monopoly and competition. In both cases illustrated there, a competitive industry will have all plants operating at $q = q^*$ in long-run equilibrium. The long-run marginal cost is to be understood as the industry's long-run marginal cost; the industry supply curve will be horizontal at this value. Figure IV.4 illustrates the determination of the number of plants. When the value of LRMC is substituted into the industry's demand equation, competitive output, Q_C is determined. Division of Q_C by q^* yields the equilibrium number of plants, N. Figure IV.4 illustrates the fact that the monopoly will produce less than the competitive industry, even in the long run, and charge a higher price for its output.

Division of Q_C by q^* will usually not yield an integer. If the number of plants must be an integer, the proceedure is quite simple. If the analysis above yields N=34.8, this means that 34 plants would each make a small profit, but 35 plants would lose money. Then the industry is in equilibrium with 34 plants present, as there is then no incentive to enter or leave the industry.

It must be emphasized that the nature of our long-run solutions depends rather heavily upon our assumptions. The most critical of these is that a monopolist owning N plants does not incur any additional costs or realize any cost savings because he operates more than one plant. (By way of contrast, the assumption that there exists an optimal plant is not overly important, since if there does not exist an optimal plant, N need not be integer). The analytical devices we have employed can be applied to more complex situations, however, and if the reader has gained some insight into their use, this chapter has been successful.

References

The following references contain detailed discussion of the maximization techniques presented and employed in this section. The list is arranged roughly in increasing order of difficulty.

1. W. J. Baumol, Economic Theory and Operations Analysis Second Edition. Englewood Cliffs, N.J.: Prentice-Hall, 1965. Chapter 1 - 8

2. A.C. Enthoven, "The simple Mathematics of Maximization," Appendix to C.J.Hitch and R.N.McKean, The Economics of Defense in the Nuclear Age. New York: Atheneum, 1965.

3. R.G.D. Allen, Mathematical Analysis for Economists. London: Macmillan, 1962. Chapter 14 and 15.

4. T.H.Naylor and J.M. Vernon, Microeconomics and Decision Models of the Firm. New York: Harcourt, Brace and World, 1969. Chapters 2 and 6 and the Mathematical Appendix to Chapter 2.

5. G. Hadley, Nonlinear and Dynamic Programming. Reading, Mass.: Addison-Wesley, 1964. Chapter 1 - 3 and 6.

6. K.J. Arrow and A. C. Enthoven, "Quasi-Concave Programming" _Econometrics_, 29 (October, 1961).

The economic concepts explored in this chapter are really fairly elementary; virtually all intermediate level price theory texts develop them fully.

CHAPTER V

The Forecasting Problems

Let us extend the scenario outlined at the start of Chapter III. Your industry was characterized by very low seller concentration in the past. Each plant was and is owned by a separate firm. Collusion to fix price and output is illegal, and there is not even a very effective trade association. The government is going to remove its controls from this market, and part of your job is to forecast what will happen if the industry remains competitive.

There is a small chance that the plants currently in the industry will be purchased by a single firm before the government controls are removed. Your employer - who is considering buying the existing capacity or sueing the firm that does buy it depending on your taste - is quite interested in what will happen under this circumstance. You as a citizen are interested in the loss society would suffer if a monpoly were formed.

The next section will describe both the data you will have available to perform these analyses and the methods to be used. We then discuss the problems in some detail. The chapter concludes with a brief examination of the sort of report that should be written about your work.

Data and Methods

In the second of the exercises described in Chapter III, you estimated the cost function of a typical plant. Total cost depended on output, factor prices, and perhaps other variables. The values of all variables exogenous to the industry will be given for this exercies, so it should be quite simple to convert your estimate into a function for total cost that depends only on output. This cost function applies to each plant in the industry, regardless of the number of plants present. In th notation of the previous chapter, you should be able to convert your cost estimate into $C(q)$.

You will also be given the values for all exogenous variables that appear in your estimated demand equation. Substituting these in, you should obtain $q = q^d(P)$, in the notation of the last chapter. You will be told the number of plants, N_0, historically and currently present in the industry. Then your estimate of the industry demand curve should be

$$Q = N_0 \, q^d(P).$$

This function describes the demand for the industry's output regardless of how that output is produced.

Thus the values of the exogenous variables appearing in your demand and cost estimates plus the value of N_0 should enable you to obtain the functions used in the last chapter's analysis. The tools and methods developed there can be directly applied to the problems discussed below.

Most of these problems involve the solution of simultaneous equations. If you have estimated simple cost and demand curves, this should be no great chore. In many cases , the function that provides the best fit to the data is a good deal more complex than another equation which fits almost as well. Where the simpler function makes the algebra easier, use it. If analytical solution is difficult for all sensible demand and cost functions available, you might investigate the possibility of using one of the standard computer programs for the solution of sets of non-linear equations. When such methods are not available, well-drawn graphical solutions are acceptable. Some instructors give credit on some problems if the equations to be solved or the computations to be performed are clearly presented, as it may be difficult to obtain numerical answers.

The Problems

Your first problem is to forecast the short-run equilibrium market price and the corresponding output (total and per plant) in your industry, given the values taken on by the exogenous variables. (Remember that each plant can produce a maximum of 100,000 units.) Then evaluate profits earned by each plant and deduce whether or not the competitive industry would be in long-run equilibrium under these conditions.

If the industry is not in long-run equilibrium, would you expect firms to enter or leave it? You should compute the long-run equilibrium number of plants in the industry. (Note that the industry is in equilibrium with N plants if that number can earn a small excess profit but N + 1 plants would suffer losses.) What are the equilibrium values of price and output?

Now suppose that all of the plants are now owned by a single firm. Forecast the price that this monpolist will charge in the short run and the among he will produce in an average plant. Evaluate total profits and profits per plant.

You should consider whether the monopolist is in long-run equilibrium. Would he make more money with more or fewer plants? If the monopolist is not in long-run equilibrium, calculate the number of plants that would be present in long-run equilibrium. Compute the long-run equilibrium price, profit and output of a monpolist. Total profit and output, as well as profit and output per plant, are of interest.

Two additional problems may be solved for extra credit. There is some risk associated with the monopolist's profits, as both demand and cost curves are subject to random shocks (error terms). The standard deviation of these variables are estimated by the standard errors of your estimated cost and demand equations. Can you estimate the variance of profit per plant in the case of short-run monopoly equilibrium?

In both short-run and long-run equilibrium, you should find that the monopolist will earn larger profits than a competitive industry. If the only effect of monopoly were to transfer resources from consumers to the monopolist, the case against monopoly would have to rest entirely on its effects on the distribution of income. Most price theory texts make it clear, though, that besides affecting the distribution of income, monopoly imposes a net loss on society - even when the monopolist is included in society. No system of unregulated markets can function efficiently is monopoly is present.

The proof of the proposition just stated can be carried out under very weak assumptions. If one wants to obtain a quantitative measure of welfare loss, however, quite strong assumptions must generally be made. Suppose that for all consumers both the income elasticity of demand for a particular good and the fraction of income spent on that good are small Then if the ith consumer purchases a quantity q^i at a price P, the area under his demand curve from the q=0 axis to $q=q^i$ can serve as an approximate dollar measure of the addition to his total utility from consuming the good in question. If society values the income of all its members equally, the corresponding area under the demand curve for an industry can be used as a measure of community welfare. This analysis is approximately correct if some sales are made to perfectly competitive consumer good industries, rather than directly to consumers.

For the purposes of this exercise, assume that the above assumptions are sensible for this industry. Figure V.1 then illustrates the computation of the welfare loss due to monopoly. A monopolist will charge a price P_M and sell a quantity Q_M. The total area under the industry demand curve from Q=0 axis to $Q = Q_M$ is the sum of the areas labeled A, B, C, D, and F. The summ of C+D+F is paid to the monopolist, while A+B is the so-called net consumer's surplus. The total cost of producing Q_M is equal to F+FC, where FC is the monopolist's fixed costs, if any. Since the cost of production represents use of society's resources, the net welfare associated with monopoly is simply

Figure V.1

Welfare Loss Due to Monopoly

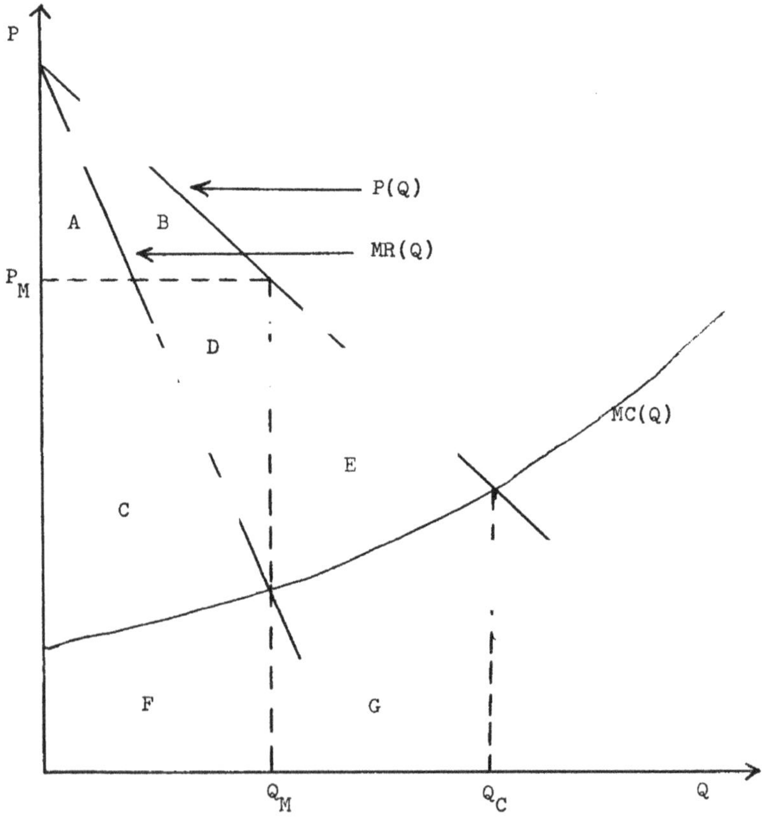

$A + B + C + D - FC.$

Under competition an amount Q_C will be produced. The same analysis that we just worked through indicates that the net welfare associated with the competitive case is

$$A + B + C + D + E - FC$$

Subtracting, the net welfare loss due to monopoly is the area labeled E. Note that this is a net loss; we've considered the monopolist on a par with all other members of society. The simplest way to compute E is to notice that it is equal to the area under P(Q) from Q_M to Q_C minus the additional cost incurred in order to produce Q_C.

As an interested citizen (and for extra credit) you might use the technique just presented to compute the welfare loss due to monopoly in your industry in the short run and in the long run.

Your Report

As we mentioned in Chapter III, a good research report moves from assumptions to method to results. That outline should be followed here.

You should begin by presenting estimated cost and demand functions you will work with. Substitute in the values of the exogenous variables and of N_0 in order to obtain the functions appropriate for this exercise. You should then outline the methods used to obtain the forecasts of short-run and long run equilibria and present your numerical results. The arithmetic should be confined to an appendix. Method is more important than results, but incorrect arithmetic here can cause you trouble in the decision-making exercise . Finally, you may wish to answer or attempt to answer either or both of the extra credit problems.

Pricing in Oligopolistic Markets

Introduction

The simplest definition of oligopoly comes from the Greek roots that compose the word. These translate simply as few sellers. Of course, "few" is one of these relative words that should never appear in a good definition, for the question "How few?" immediately arises. The answer in this case is as follows: an oligopoly is an industry with few sellers, so few that each is aware that his profits depend on the decisions of each other firm inthe industry. In shorter form, an oligopoly is a small group of firms, selling similar products, where each firm recognize the impact of each other firm on his profits.

The distinguishing feature of oligopoly is that there is no neat relation between industry and firm demand curves. An oligopolist knows that it is likely that his rivals will, at least eventually, become aware of any price change he institutes. The final change in sales he can expect to result from a price change depends on his rivals' reactions to the change. Since these reactions cannot be known in advance, each firm's decisions will be based on the response it expects to elicit from the other firms in the industry. It is the basic importance of expectations that makes it difficult to draw precise careful analyses of industry behavior by empirically-oriented economists as well as on deductive theory. The references at the end of the chapter should be consulted by the interested student.

The bulk of our discussion will focus on price competition. An important feature of industries with few sellers is often the heavy use of the tools of non-price competition, especially advertising. We shall examine advertising briefly in the last section of this chapter. Our discussion there may be anticipated by the general remark that many of the things we shall say about price competition are also true of non-price competition.

The next section considers the oligopoly problem in the static case. We assume that firms set price once and only once. We then bring in dynamic considerations in the following section. There we examine the more realistic situation where each firm may change its price at any time. These two sections assume no entry into the industry is possible; we then relax this assumption. The final section summarizes our conclusions and briefly considers non-price competition.

The Static Problem

The essence of the problem facing oligopolists is captured by the so-called Prisoner's Dilemma game. The situation considered is as follows. Two mean are arrested for robbery. Each is given the spend a year in prison on a minor charge. If both testify, both will spend ten years in prison. If one testifies and the other doesn't, the one that says nothing will be sentenced to twenty years and the informant will go free.

It is clearly in the interest of both prisoners that neither testify. But if they are not allowed to talk with each other, it is very likely that both will testify. By testifying each can insure that he will spend no more than ten years in prison, while if either refuses to talk, he may end up in jail for twice as long.

A concentrated market situation bears a striking resemblance to this simple game. If all firms charge high prices, all will generally earn a high profit. If all charge low prices, they will all earn little money. This fact alone would suggest that high prices would always prevail. But the situation is unstable in the same way the prisoner's dilemma is: if all firms but one charge high prices, and that one firm undercuts the others, he will generally earn a very large profit. Thus all firms have a strong incentive to persuade their rivals to charge high prices; the incentive to follow them is weaker.

We thus identify two basic tendencies in oligopolies. The first is the tendency for all firms to try to band together to raise prices and to keep them high, to <u>collude</u> explicitly or implicitly and form a <u>cartel</u>. If such a policy is successful, total profits for the group of firms can be maximized. The second tendency results from the fact that any firm can maximze its individual profits by cheating on an agreement to charge high prices, so that any such agreement may very well be unstable. The basic concern of this chapter is to identify and discuss factors affecting cartel formation and stability.

A simple example will illustrate the dilemma faced by firms in concentrated markets. Suppose there are two firms, selling amounts Q_1 and Q_2 of identical products.

The industry price, P, is determined by

$$(6.1)\quad P = A - B(Q_1 + Q_2)$$

For simplicity, we assume that marginal cost is zero. If we let total industry output be Q, it is clear that revenue is given by

$$(6.2)\quad R = AQ - BQ^2.$$

Revenue, and henceprofits, will be maximized if $Q = A/2B$. Thus for maximum industry profits, the firms should agree to set $Q_1 = Q_2 = A/4B$. Each firm will then earn a profit of $A^2/8B$. Suppose that firm one adheres faithfull to an agreement to sell A/4B units. If firm 2 knows that firm one will do this, its revenue is given by

$$(6.3)\quad R_2 = AQ_2 - B(A/4B + Q_2)Q_2.$$

The revenue of firm 2 can easily be seen to be maximized by $Q_2 = 3A/8B$. This will yield a market price of 3A/8. Firm two's revenue will be $9A^2/64B$, and firm one will earn $3A^2/32B$, two-thirds of firm two's earnings. Total industry profits will be less than they would have been under a cartel, but firm two will be better off. (Total profits may decline further, and firm two's profits may fall below the cartel level, if firm one retaliates. We shall return to this point below.)

Ignoring the possibility of price change and of outsiders entering the industry, let us consider some general factors affecting the formation and stability of cartels. (A situation where these assumptions are not unreasonably restrictive might be a group of firms submittting sealed bids on a large government project.) The first consideration is the <u>legal structure</u> the industry is operating under. If an agreement to charge high prices can be written up as an enforceable contract, it will be more stable than if this is not possible. This guarantee of stability, in turn, gives individual firms a greater incentive to enter the agreement, since the likelihood that they will be cheated on is reduced. Another consideration is whether profit-sharing schemes can be drawn up and enforced. If this possibility exists, for example, firms can decide which one of them will be the low bidder on a project and then agree to share the profits earned by this company. Under the English common law, agreement to fix prices or to share profits (or markets) are not illegal, but they cannot be enforced in court. The likelihood of a successful cartel is reduced when on price policies, and no legal sanctions can be brought to bear on cheaters. Thus the American legal structure works against the formation of stable cartels both by limiting the power of a cartel to maintain discipline and by restricting communications among cartel members. Under the common law, firms could meet to discuss and agree on prices; this is quite illegal under the Sherman Act. As a general rule, anything which amkes communication among actual or potential cartel members more difficult works against a stable cartel agreement. The greater the number of firms in an industry, for instance, the more difficult it will be to organize a cartel agreement.

Another influence is the similarity of the firms. If all firms are identical, and all have the same expectations about the future, it will be easy to agree on the strategy which will provide each firm with identical profits. If firms' cost structures differ, however, the strategy that maximizes total industry profits may leave some firms badly off. (For instance, it may be optimal for all production to be done by the most efficient company). In this case, it will be necessary for the firms that stand to gain the most to compensate those that would be harmed by the agreement before the latter can be induced to often called, are permitted by law (the legal structure enters again), it may be difficult to agree on the amounts to be paid. If side payments are illegal, it becomes even more diffiault to obtain and sustain agreement among firms with different costs structures.

Another aspect of similarity among firms is how they view the industry's environment. I fall firms have the same ideas about the industry's demand curve and the supply curves of the factors of production it uses, they will find it easy to agree on the strategy that maximizes their total profits. If their views of the world differ drastically, however, it may be impossible to form a cartel, simply because the potential members cannot agree on its optimal policy.

Finally, we must consider the <u>incentive for cheating</u>, on an explicit and implicit cartel agreement. Both cost and demand conditions enter here. The more responsive <u>demand</u> is to price differences within the industry, the greater will be the incentive to cheat on any price-fixing agreement. Generally, if the products of a group of sellers are viewed as nearly identical by buyers, each firm will find that the demand for its output is very responsive to the price it cahrges. Under such conditions, the incentive to cheat on a cartel can be quite enormous. This makes for instability of any agreement, but it may cause an agreement to be formed in the first place. If the firms each stand to lose a great deal if they are undercut, they may repeatedly try to set up mechanisms that will make underautting less likely. This is especially true if industry demand is inelastic, so that the difference between cartel and competitive profits is large. On the other hand, to the extent that products are differentiated in the minds of buyers, the vulnerability of a cartel to cheating is reduced, but the incentive to form an explicit cartel in the first place may be also lessened. The usual expectation is that, since the incentive to cheat is less when products are differentiated, an industry is more likely to have a stable cartel in this case. The cartel will very likely be implicit.

The impact of product differentiation on the incentive to cheat can be illustrated algebraically. Consider a market with two sellers. The firms' products are not viewed as identical by consumers. They set prices P_1 and P_2, and sales are determined by the demand relations

$$Q_1 = a + kP_2 - (b+k)P_1, \quad \text{and}$$

(6.4)

$$Q_2 = a + kP_2 - (b+k)P_2$$

Industry demand when both firms charge a price P is given by

$$(6.5) \quad (Q_1 + Q_2) = 2a - 2bP$$

The larger is k, the more sensitive the industry's customers are to price differences within the industry. When k is very large, the two firms are producing products which the industry's customers consider essentially identical. If k=0, both firms are pure monopolies.

Assume, for simplicity, that both firms have zero marginal costs. Then total profits will be maximized when $P_1 = P_2 = a/2b$. If both firms change this price, both will earn a profit of $a^2/4b$.

Suppose firm one charges a price of a/2b, and firm two feels that his competitor will maintain that price no matter what. Then firm two's profit can be written as

$$(6.6) \quad R_2 = (a + ak/2b)\, P_2 - (b+k)\, (P_2)^2.$$

This quantity is maximized by $P_2 = (a/2b)[(2b+k)/(2b+2k)]$. Notice that for k=0, there is no incentive to cheat. The larger is k, the lower is the optimal cheating price. Substitution of the expresssion for P_2 into (6.6), we find that the cheater's profit is equal to

$$(6.7) \quad [\frac{a}{4b}]^2 \quad \frac{(2b+k)^2}{(b+k)}$$

Expression (6.7) rises when k is increased. Thus the less differentiated the products, the greater the incentive to cheat.

<u>Cost conditions</u> also affect the incentives to cheat on an implicit or explicit cartel agreement. The more rapidly marginal cost rises, the less the incentive to lower price to acquire additional sales. similarly, the more short-run excess capacity a firm has, the greater its incentive to cheat.

and cost conditions. We shall next relax the assumption that firms' prices (and conditions in the industry in general) do not change. Additional forces affecting cartel stability will be suggested, and additional effects of the elements discussed above will be examined.

The Dynamic Problem

We still retain the assumption that no outside firm can enter the industry. Further, to have a situation worth discussion, we shall assume that the firms are not parties to an enforeable cartel agreement. The new element in our analysis is the fact that demand conditions and prices may change over time, and firms' strategies must take this into account.

The problem of finding an optimal cartel strategy is now replaced by the problem of first finding such a strategy and then modifying it as circumstances change. Thus anything that makes it difficult to decide on an optimal strategy will make it even more difficult for the cartel to respond to changes in its environment, since effecting such responsees amounts to renegotiating the cartel's policy. Legal structures or number of firms that make face-to-face negotiation difficult will lessen the ability of the cartel to function, and a cartel that cannot react to changes in its environment dominant firm. Then the dominant firm is the natural leader, and it is easy to reach agreement to follow its lead. It may be difficult to sustain stability, though, since it may be hard for the dominant firm to detect cheating. Also any effort on its part to punish cheaters may lead to the collapse of the cartel agreement.

The major change in the model when we bring in dynamic elements is the nature of cheating on an explicit or implicit agreement. Given the ability of the faithful cartel members to alter their prices, it becomes likely that a cheater will eventually face retaliations. Besides considering the short-run sales and profits gained by undercutting the cartel, a potential cheater must consider the impact of the cartel's retaliation. If his action depresses everyone's profits for a long time to come, this may outweigh his short-run gains. A cheater thus must make decisions on the basis of his expectations of the form of the cartel's response to his action. If the cartel does nothing, he will continue to earn veyr high profits. If the cartel merely matches his low price, cheating will still usually be profitable. If the cartel reacts by undercutting the cheater, however, cheating can be rendered unprofitable.

is the time until it occurs, the cartel's reactions lag. This can be divided into two intervals: the time it takes to detect cheating and the time it takes to effect a response. The longer the cheating can go undetected, the more profit the cheater will make. Similarly, the longer it takes the faithful cartel members to respond to his action once they're aware of it, the more profitable it is to cheat. We have spoken of the cartel's decision process above. The new element is the problem of detection of cheating behavior.

Before we discuss the detection of cheating, it will be useful to examine the dynamic problem facing a potential cheater algebraically, We suppose that r is the rate of interest used to discount future earnings, and we further assume that interest is compounded continuously. Then it is easy to show that one dollar invested today will build into e^{rt} dolars in t periods, where "e" is the base of the system of Naperian or natural logarithms. Then one dollar to be paid t periods in the future is worth e^{-rt} dollars today. We shall write this expression as exp(-rt). When either r or t is zero, this expression equals one, and it declines to zero as rt becomes large.

Suppose that the potential cheater could earn profits of R^L per priod if he remained loyal to the cartel. If he cheats, he expects to earn profits of R^e until his cheating is detected. He expects to be detected after τ periods. At this point the cartel will retaliate, and then he expects to earn R^W("W" for warfare) for T periods. At the end of

R^L. Similarly, it is never profitable to cheat if $\tau = 0$ or if T is very large.

Now suppose that r is non-zero, that one dollar today is indeed worth more than a dollar tomorrow. Recall from integral calculus that

$$\int_{T_1}^{T_2} \exp(-rt)\, dt = (1/r)[\exp(-rT_1) - \exp(-rT_2)].$$

The <u>present discounted value</u> of the profits of a loyal cartel member is given by

$$(6.9) \quad PV^L = \int_0^{T+\tau} R^L \exp(-rt)\, dt = (R^L/r)[1 - \exp(-r(T + \tau)].$$

Similarly, the present value of a cheater's profits is

ratio of these two quantities may be written

$$(6.11)\quad PV^C/PV^L = (R^C/R^L)F + (R^W/R^L)\ (1 - F), \text{ where}$$

$$F = [1 - \exp(-r\tau)] \;/\; [1-\exp(-r(T+\tau))].$$

Thus (6.11) is a weighted average of the same sort as (6.8).

The larger is F in (6.11), the more likely it is that cheating is profitable. When $\tau = 0$, when detection is instant, F is zero, and cheating is never profitable. When r is very large, a dollar earned today is worth much more than a dollar lost tomorrow, and cheating become profitable. When T is zero, cheating is also always profitable. When T is very large, (6.11) reduces to

$$(6.12)\quad PV^C/PV^L = (R^C/R^L)[1-\exp(-r\tau)] + (R^W/R^L)[\exp(-r\tau)].$$

For any r, it will always be possible to find a τ small enough to make cheating unprofitable.

Notice that all the quantities in (6.8), (6.11) and (6.12) except for r and R^L are values <u>expected</u> by the potential cheater. The problem of a cartel wishing to deter cheating is thus to influence the <u>expectations</u>

of its members about the values of the paremters that determine the profitability of breaking the cartel agreement. If all firms feel that cheating would not be profitable, none will cheat, even though all may be wrong.

We now consider the information availabel to each firm that can be used to detect cheating. If each knows the others' prices, detectio of cheating is easy. A trade association that is legally empowered to obtain and publish prices, for instance, makes for cartel stability. A voluntary association is, of course, vulnerable to false price reports by cheating firms. If each firm can detect cheating only by analyzing changes in its own sales, the detection problem is more diffiuclt. The firm must make a judgement about which sales changes are attributable to the myriad random elements determining demand and which are due to a rival's cheating. If the number of firms is large, it will usually be unable to do this with any confidence. Indeed, usual rationale for the model of perfect competition: each firm is small that it is unable to influence noticeably any individual competitor. Similarly, no firm can, by itself, detect any alteration in the policies of any other single company. No cartel is stable under these conditions, as many farmers' organizations have found to their sorrow.

All other things being equal, it is clear that the more predictable are individual firms' demands, the easier it is to detect cheating. The more "noise" there is in the market, the harder it will be to detect a cheater.

cheater, a firm may elect to remain faithful simply because that is the less risky course.

Once cheating is detected, subsequant behavior can be quite interesting. We shall present two special caes which have the advantage of illustrating the sort of non-collusive equilibria often encountered in theoretical discussion of oligopoly.

Consider first the duopoly (oligopoly with two firms) selling identical products whose demand structure is described by equation (6.1). Below that equation we discussed the case where firm two cheated on a cartel agreement, assuming that firm one would not respond. Suppose now that firm one does respond. It can take a variety of courses: a sensible move might be to maximize its profits given firm two's output. If it does this, will produce and sell 5A/16B, undercutting firm two. If firm two retaliates in the same fashion and the process continues, equilibrium will be reached at the point where both firms are producing A/3B. This point is called the Cournot equilibrium point. It is stable change in output. But, of course, both would be better off if they could re-establish $Q_1 = Q_2 = A/4B$. It should be noted that the Cournot equilibrium model can be easily extended to the case where there are more than two firms in the market.

We shall quickly show that the point $Q_1 = Q_2 = A/3B$ is indeed a non-collusive equilibrium in this model. Consider the problem of maximizing the profit of firm one, subject to the given output of firm two. To solve this, we must maximize

$$(6.13)\quad AQ_1 - B(Q_1+Q_2)Q_1,$$

treating Q_2 as a constant. Differentiating (6.13) with respect to Q_1 and setting the derivative equal to zero, we obtain

$$(6.14)\quad A - B\,Q_2 - 2\,B\,Q_1 = 0.$$

Equation (6.14) gives Q_1 as a function of Q_2. Suppose that both firms are selling the the same amount and that equation (6.14) is satisfied. Then neither firm has an incentive to alter its production, and the point is an equilibrium. Setting $Q_1 = Q_2 = Q$ in (6.14) and solving, we obtain $Q = A/3B$.

Now consider duopoly with differentiated products, as described by demand schedules (6.4). Total profits are maximized when $P_1 = P_2 = a/2b$. If firm one if faithful to an agreement to charge this price, firm two will maximize its profits by setting

$P_2 = (a/2b)[(2b+k)/(2b+2k)]$. When it detects cheating, firm one may retaliate by maximizing its profits given the price charged by firm two. It will then set a price below firm two's.If firm two then responds in the same fashion, the process will continue until prices have fallen to the point where $P_1 = P_2 = a/(2b+k)$. At this point, neither firm has an incentive to power price further. Again, we would assume that both would seek to restore a price of $a/2b$.

Notice that when k is very large, the price cutting does not stop until price is zero (or until price is equal to marginal cost, when marginal cost is positive). This result of price-setting is associated with the names of Edgeworth and Bertrand.

To provide that $P_1 = P_2 = a/(2b+k)$ is really a non-collusive equilibrium, we proceed as above. Consider the problem of maximizing the profits of firm one, given the price of firm two. Firm one will then be maximizing

$$(6.15) \qquad aP_1 + k\, P_1P_2 - (b+k)(P_1)^2$$

with respect to P_1. Setting the first derivative equal to zero, we obtain

$$(6.16) \qquad a + k\, P_2 - 2(b+k)P_1 = 0.$$

In equilibrium, it must be the case that $P_1 = P_2$ and this equation is satisfied. The neither firm has an incentive to change its output. Making this substitution and solving, we obtain the result stated above.

cartel - cheater dichotomy may be a bit unrealistic. A firm may alter price as a signal to other firms about where it thinks the cartel price should be. The others must, somehow, distinguish this from cheating. As we saw above, generally the incentive to cheat rises as profits rise, and a cartel may find it best not to attempt to maximize total industry profits. It may instead maintain price at a level below the total profit-maximizing level in order to reduce the incentive to cheat and thus to increase stability. The level at which price stablizes will then depend on the mutual trust of the various firms, as well as on their attitutdes toward and estimates of risk. A cheater may, by his action, lower the level of trust to a point where it is impossible for the cartel ever to attain maximum total profits.

The above analysis has considered oligopoly behavior as a dynamic problem in the face of risk. Attitudes toward risk were seen to affect firms' stategies. We saw that the rate at which future profits are discounted will information available to firms, especially as regards their ability to detect cheating, the ability of the cartel to respond to cheating and the form of that response, and the level of trust in the industry. So far, we have taken the firms in the industry as given. It is now necessary to relax this assumption also.

The Impact of Potential Entry

The assumption of the perfectly competitive model is that firms can enter and leave industries instantly and costlessly. It is further assumed that firms will enter industries where excess profits are being earned and leave lines of business where profits fail to cover opportunity cost. The textbook monopoly model begins by assuming that no firms can enter the monopolist's industry, no matter what he does. As one might expect, conditions of entry in the real world are usually somewhere between these extremes. There are barriers to entry, but these are not usually so great as to completely exclude new firms from entering any industry.

The literature distinguishes three basic kinds of barriers. The first is absolute cost differences between new and established firms. The established firms may have a critical patent or important know-how, or they may control the most desirable raw material sources. It is often alleged that the capital market favors established firms, so that new entrants must pay a higher price for funds.

relative size. The larger the minimum efficient firm size relative to a particular market, the less easily an entrant can sneak unnoticed into that market, the more likely his entry is to upset prevailing patterns of price formation. If, for instance, a firm need only produce 1% of the current total output of a particular industry in order to be efficient, he can probably enter the industry safely assuming that he will not be noticed. His entry is unlikely to cause a fall in price. On the other hand, if a firm would need to produce 50% of current industry sales in order to be efficient, his entry would add 50% to industry output. Prices would fall, and the final equilibrium of the industry would be uncertain.

Finally, the literature often speaks of product differentiation barriers to entry. The stronger are the preferences of buyers for the products of existing firms, the more an entrant is at a disadvantage. If he can change these tastes with a massive advertising campaign, the necessity of carrying out such a campaign constitutes an absolute cost barrier to entry.

The best way to consider entry is as a stochastic process. The probability of entry in any time period will vary inversely with the level of barriers to entry and directly with the profits being earned by the established firms. It will also depend heavily on how potential entrants think the established firms would react to entry; the problem of deterring entry is very much like the problem of preventing cheating. If potential entrants feel the existing firms would not change their price and output policies if entry occurred, entry is more likely On the other hand, if the established firms can convince potential entrants that their entry would send prices tumbling,, entry is less likely. Given technical entry barriers and the expectations of potential entrants, there will generally be some level of excess profits greater than zero that can be earned by the established firms that will make the probability of entry zero.

How does the possibility of entry affect oligopoly behavior? It seems likely that the impact on cartel stability will generally not be great. Conditions of entry must be considered by the cartel, of couse, and to the extent that cartel members disagree about how to do this, the cartel is less stable. But it is not likely that the height of entry barriers will affect the cartel's stability.

The main impact of finite barriers to entry would seem to be on the optimal cartel strategy. If a cartel seeks to maximize total industry profits, it must consider the possibility that a strategy that maximizes profits in the short run will induce entry and thereby result in lower profits in the long run. As before, short-run gains must be weighted against the possibility of long-run losses. In general, the lower are entry barriers, the lower will be the cartel's optimum price.

Summary and Conclusions

So far we have explicitly considered only price competition It can be shown that the same basic conceptual framework applies to other forms of competition as well. Consider, for instance, advertising. It is generally the case in established industries that if all firms can agree to agree on he level of advertising, they can choose this level so as to maximize industry profits. If they all double advertising, all will make less money. But if all firms but one are adhering to the cartel advertising policy, and one firm is advettising more than its quota, the cheater stands to gain. We have the prisoner's dilemma again, and the rest of our discussion above can be applied more or less directly to non-price competition.

The main difference between price and non-price competition is that agreements to limit non-price competition are generally quite difficult to formulate and enforce. There are so many forms of non-price competition that if one, say the number of salesmen, is controlled, firms will often cheat on the spirit of the agreement by using another form, say advertising , more intensively. In addition, it is more difficult to detect cheating on the letter of an agreement on non-price competition. The impact of advertising on sales is not generally easy to predict, and firms do not normally publish their advertising budgets. It would thus be very difficult to detect violations of an agreement to limit advertising spending. In some cases these problems can be circumvented by agreements directly assigning individual buyers to individual sellers, but this is not usually possible. (It is illegal in this country). As a consequence successful collusion to limit non-price compeition is quite rare.

We can summarize the analysis of this chapter by grouping the factors we have considered under three broad headings. First are those that affect the ability of firms in an industry to agree on the policy that will maximize their total profits, the optimal cartel strategy. SEcond, we shall summarize those forces that affect the ability of a certel to obtain high profits. Finally, we shall mention those factors that influence cartel stability.

The greater the degree of communication among the firms in an industry, the greater the likelihood that they can agree on a cartel strategy. Communication possibilities will be affected by, among other things, the legal structure and the number of firms involved. The more nearly identical firms' cost structures and views of their environment, the easier it will be to arrive at an agreement. Finally, the simpler a good cartel agreement can be, the more likely one is to be arrived at. If all firms sell one product at one price, agreement will be easier than if each sells a variety of products in a variety of locations at a variety of prices.

The lower are entry barriers and the more strongly would-be entrants expect the cartel to react to their entry, the greater are potential cartel profits. The less likely high profits are to attract antitrust prosectution , the more likely it is that a cartel will be able to maximize industry profits. The more the members of an industry trust each other, the more likely it is that they will be willing to charge the price that maximizes their total profits, since this price will render cheating very profitable. Two general considerations interact with all these elements. The more tolerant the members of the industry are to risk, the more likely it is that they will seek the highest possible profits. Finally, the more heavily future profits are discounted (that is, the higher the relevant interest rate), the more likely it is that a cartel will pursue a strategy that maximizes current profits.

Cartel stability is affected by a wide variety of forces. The more sensitive demand is to price differentials within an industry, the flatter are marginal cost curves, and the greater is excess capacity, the greater will be the immediate profits earned by a cheater. The more effectively each firm expects the faithful cartel members to respond to his cheating, and the quicker they can do so, the less the incentive to cheat. The speed of cartel response will be positively related to the amount of information availabel to each firm, relavant to detecting cheating. All other things being eqal, it will be easier to detect cheating the fewer the firms in be industry and the more stable each firm's demand. Since cheating is usually risky, the less tolerant firms are of risk, the less likely they are to cheat. The more heavily they discount future profits as opposed to present profits, the more likely they are to cheat. Finally, the less tolerant the legal system is of formal cartel arrangements, the more likely is cheating.

It should be obvious that a large number of forces influence conduct in oligopolistic industries. The net impact of any one on industry profits may be difficult to ascertain. We do have some insights into how concentrated industries perform, although these observations by no means constitute a full-developed theory.

References

As many texts do not cover the topic of oligopoly behavior in much depth, we present here a fairly complete set of references. Not all are theoretical works, it should be noted. The list is in alphabetical order, as no other arrangement seemed much superior.

1. J.S. Bain, *Barriers to New Competition* (Cambridge: Harvard University Press, 1956). Especially chapters 1 - 5.

2. R.L. Biship, "Duopoly: Collusion or Warfare," *American Economic Review*, 50 (September, 1960).

3. E.H. Chamberlin, *The Theory of Monopolistic Competition* 8 th edition, (Cambridge: Harvard University Press, 1962). Especially chapters 3-5.

4. W. Fellner, *Competition Among the Few*, (New York: Knopf, 1949). Especially chapters 1 and 7.

5. R. B. Heflebower and G. W. Stocking, editors, for the American Economic Association, *Readings in Industrial Organization and Public Policy* (Homewood, Illinois: R.D.Irwin 1958). Especially chapters 4 - 7, 9, 10, 12, and 17.

6. F. Modigliani, " New Developments on the Oligopoly Front" , *Journal of Poliitical Economy*, 66 (June, 1958).

7. D. Orr and P. W. MacAvoy, "Price Strategies to Promote Cartel Stability", *Economica*, 32 (May, 1966).

8. G.J. Stigler and K.E. Boulding, editors, for the American Economic Association, *Readings in Price Theory* (Homewood, Illinois: R.D. Irwin, 1952). Especially, chapters 18 and 20 - 22.

9. G.J. Stigler, *The Organization of Industry*, (Homewood, Illinois: R.D.Irwin, 1968). Especially chapters 5,16 and 18.

This chapter outlines an exerciese primarily designed to illustrate the problems and results of price-making in concentrated markets. The computer program is by no means limited to consideration of such ceses, though. Competition and monpoly can be equally easily simulated.

You will be divided into teams which function as firms. Each firm begins in a particular industry composed of a number of iden'ical competitors. The sales and profits of each firm are affected by the prices charged by each one of its competitors. (In the event that more than one industry is simulated, prices in any one industry have no effect on the other industries.) The objective during the "play of the game" is to maximize profits; after the last computer run you will be asked to report on the results of the exercise. The next section outlines the structure of the simulated markets. We then discuss the play of the game, and we conclude with a few remarks about your report on the exercise.

Market Structure

Every firm has control over one plant of the type whose cost function you have already estimated. The values of the exogenous variables are the same here as in Chapter V, so the cost curve C(q) used for the forecasting exercise is the same one that describes each firm here. Remember that there is an upper limit of 100,000 units on output per period.

Since the values of all exogenous variables are the same as in the forecasting exercise, the plant demand curve $q^d(P)$ you obtained there carries over also. This will generally be the demand curve facing each firm in this exercise if all firms charge the same price. (The instructor may raise or lower this curve by multiplying quantity demanded by a constant; you will be told if this has been done.) What happens when firms charge different prices? Industry demand will generally depend on the unweighted average of the prices of the firms present, while firms' demands will depend on the differences between their price and the industry average. The estimation exercise provides no information on the sensitivity of firm demand to price differences within industries; you will have to learn about this as the exercise progresses. This sensitivity may vary from industry to industry.

You may well want to use the observations on cost and demand that are generated during the game to sharpen your estimates of the cost function and of the industry demand function. Both cost and demand functions still contain error terms.

Firms are assinged four-digit numbers for identification purposes. The first two digits of this number refer to the industry in which the firm originated. For instance, firm three in industry four would be assigned the number 0403, while firm four in the same industry would be designated 0404.

We shall first describe the mechanism by which individual firms' decisions are translated into the sales, costs, and profits of all firms in an industry. We then discuss four factors that vary from industry to industry: entry conditions, legality of side payments, legality of collusion, and information availability.

The demand for the output of each firm depends on all prices posted in the market in which the firm operates. If any firm's price is too high, it will receive a zero demand, and it will sell nothing.

Firms for which demand is greater than 100,000 units sell 100,000 units and have unfilled orders equal to the difference between their demand and 100,000. These firms are then put to the side and the simulation program divides their unfilled orders among those firms with excess capacity and positive demand. To do this, it first calculates the total demand in the industry. It then computes the share each of the firms with positive sales and excess capacity received of this total demand. Each then receives a fraction of total unfilled orders equal to its share of total industry demand. Thus if any firm's demand is zero, it will receive no unfilled orders. If the firms with excess capacity accounted for only half of industry demand, they will receive only half of the unfilled orders generated by the other firms. If this allocation raises some demands above 100,000 units, the unfilled orders so produced are allocated in this same fashion.

The basic idea underlying this simplemodel is that a dissatisfied retailer does not immediately withdraw his orders. He places some additional orders with companies having excess capacity (and charging higher prices), but his demand is reduced. The process continues until all orders are met or until all firms have unfilled orders. If there are no unfilled orders generated, total industry sales will obviously equal total industry demand. If any firms have unfilled orders, though, total industry sales will be less than total industry demand.

A numerical example of this allocation mechanism is shown as Table VII. Notice that total industry demand was 330,000 units. Because most of it was concentrated in the hands of firm 4, which was unable to satisfy it, though, total industry sales came to only 266,100 units. Total sales plus unfilled orders equal 370,400 - a figure with almost no meaning.

The relation of sales to demand thus depend on what happens to unfilled orders. When there are no unfilled orders in the industry, sales equal demands. When unfilled orders are present, it is often possible to compute what demands must have been to give rise to the observed distribution of sales and unfilled orders.

At the discretion of the instructor, it may be possible for outside firms to enter some industries. The entrant gives up his old plant and begins operating a plant identical to those in the industry he enters. He is, of course, subject to the demand relations that characterize his new industry. A lump-sum payment is assessed each time a firm switches industries; the amount depends on the industry entered and is fixed by the instructor. Any attempt to enter an industry into which entry is not allowed will result in the firm being restored to the industry in which it last operated. Any attempt to enter an industry with ten firms already present will be similarly frustrated, as the computer program is unable to handle industries with more than ten firms.

unchanged on average.(Remember the error terms.) The effect of entry is to divide the market among all firms present; if there is net exit the "leavings" are divided among all firms remaining behind. The standard deviation of the random (error) term of firm demand functions rises and falls inversely with the number of firms present; a smaller piece of the action means less absolute variation.

Suppose, for instance, that in period one an industry had four firms each charging $2 and that in period two another firm entered at the same price. Industry sales would (on average) be the same in the two periods, but each firm in the second period would sell (on average) only 4/5 as much as it did in the first period.

Firms operating in some industries may be permitted to make and receive lump-sum cash transfers (side payments). Payments made to a firm inelegible to receive them or to a non-existent firm will be debited to the payer but not credited to the payee. Payments made by a firm inelegible to make them will also be debited. Negative side payments will be treated as positive.

The enforceability of contracts calling for side payments will depend on the legal structure. Contracts that call only for cash payments will, in general, be enforeceable. But contracts that specify a cash payment in exchange for some service that can be held in restraint of trade - such as not entering a particular industry or charging a particular price - will be enforecable only when any contracts are enforceable. All enforcements will be handled by the instructor; he will specify the mechanics involved.

There is a dummy firm present whose number is 9999. This firm acts as the instructor's fiscal agent. It receives all entry cost payments, and it can make side payments and levy fines. In addition, it can receive payments from any firm.

The extent to which collusion is allowable in price determination must be specified, and it may vary from industry to industry. Firms will be subject to one of three legal structures:

1. The Sherman Act: All contracts and conspiracies in restraint of trade are illegal (You cannot get together to fix price, regulate entry, etc.)

2. The English Common Law: Firms may collude on price and/or agree to make side payments, but such agreements are not enforceable. (The instructor will not fine cheaters for you).

3. The (representative) German Kartell Law: Firms may collude on price and/or agree to make side payments, and any such agreements are enforceable.

Again, unless it is stated otherwise, the instructor will perform any and all judicial functions that may be required during this exercise, including interpreting the Sherman Act.

The amount of information given firms operating in each industry about conditions in other industries and about the actions of competitors may vary from industry to industry at the discretion of the instructor. At a minimum, all firms are shown their own price, sales, profit and unfilled orders. There are sixteen different possible configurations relating to the firm's competitors. The following information is available: the identities of the firms present, and their prices, sales, unfilled orders, and profits. Firms will appear in random order on output lists; identity cannot be inferred from firm number and position on output lists. INformation not available to a particular firm will generally show as a blank on its output sheet in each period.

Play of the Game

Mechanically, this exercise is quite simple. Each firm must submit a firm card, punched exactly as shown in Table VII.2, which will remain in force until a new card is submitted. Be very careful to punch data in the correct columns. A mis-punched card will stop the computer program and greatly annoy the instructor. The procedure for submission of firm cards will be announced in class.

At the end of each run, each firm will receive interim output giving the results of that period's operations. You may then elect either to change your firm card to to leave it in force. The time schedule for runs will be announced at the start of the exercise.

Conceptually, the problem of making good decisions is not at all simple. You would like all firms in your industry to charge the short-run monopoly price, which is the same here as it was in the forecasting problem, unless the instructor has multiplied firm demands by a constant which he has given you. In this latter case, the new monopoly price is not hard to compute.

Persuading other firms to charge and stick to the monopoly price will not be easy in general. Profits may often be raised dramatically in the short run by undercutting competitors. Against these gains a potential cheater - which means any firm - must weigh the losses that will be imposed by retaliation. In general, you should read and understand Chapter VI before beginning to make decisions.

The first period of play is particularly difficult in the absence of any explicit cartel agreements. You will clearly want to set a price somewhere between the short-run competitive and monpoly solutions. Where in this range you elect depends, as all your actions in this game must, on what you expect your competitors to do.

Your Report

At the end of the simulation exercise, you will receive a set of summary output, giving information on other firms and industries not provided by interim output. In your report you should analyze this data, and explain what happened in this exercise. You should make use of the ideas in the last chapter and try to relate industries performances to their structures. You should be able to formulate and (casually) test some hypotheses.

Particular attention should be paid to the firm and industry or industries in which you were involved. But a simple annual-report-style chronology of the actions of a particular firm is of little or no interest here.

Table VII.1

An Example of the Allocation of Unfilled Orders*

Firm	Original Demand	Share of Original Demand
1	10	.03
2	90	.27
3	40	.12
4	190	.58

Round 1			Round 2	
Sales	Unfilled Orders		Sales	Unfilled Orders
10	0	[10+(.03)(90)=12.7]	12.7	0
90	0	[90+(.27)(90)=114.3]	100	14.3
40	0	[40+(.12)(90)=50.8]	50.8	0
100	90		100.	90.0

Round 3**

	Sales	Unfilled Orders
[12.7+(.03)(14.3)=13.4]	13.4	0
	100.0	14.3
[50.8+(.12)(14.3)=52.7]	52.7	0
	100.0	90.0

* All figures are in thousands of units.

** Round 3 is the final round, and it is only one that shows on computer output.

Table VII.2

The Format for Firm Cards

Contents

The four digit (no decimal point) firm number--to be assigned. Consists of a two digit original industry number and a two digit firm number.

The industry the firm is to operate in this period. May be left blank if no change in industry is desired.

the price to be charged (punch a decimal point.)

The four digit firm number of the firm to which a side payment is to be made. may be left blank if no side payment is to be made.

The amount of the side payment in dollars (punch a decimal point). This may also be left blank. (If this number is negative it will be made positive by the program.)

Made in the USA
Monee, IL
09 December 2023

48620917R00080